Six Sigma Instructor Guide

Green Belt Training Made Easy

Six Sigma® Instructor Guide 2nd Edition
© 2003 by Jay Arthur

Published by LifeStar
2244 S. Olive St.
Denver, CO 80224-2518
(888) 468-1535 or (303) 281-9063 (orders only)
(888) 468-1537 or (303) 756-9144 (phone)
(888) 468-1536 or (303) 753-9675 (fax)
lifestar@rmi.net
www.qimacros.com

Upgrade Your KnowWare®!

ISBN 1-884180-22-1

All rights reserved. Permission to reprint quotes and excerpts in company or business periodicals–magazines or newsletters–is given and encouraged with the following credit line: "reprinted from *Six Sigma Instructor Guide* by Jay Arthur, (888) 468-1537."

Any Six Sigma book can be customized to reflect a company's improvement process. For information, call, write, or e-mail to the address above.

Also by Jay Arthur:
The Six Sigma Simplified, LifeStar, 2001, 1-884180-
Improving Software Quality, John Wiley & Sons, 1993, 287 pages, ISBN 0-471-57804-5

Phone, Fax, or E-mail support: Contact Jay Arthur at the phone, fax or e-mail address above for any questions you have about Breakthrough Improvement or using this book.

Jay Arthur, the KnowWare® Man, works with companies and managers who want to close the gap between where they are and where they want to be. Jay worked with Baby Bell teams that:
• Saved over $36 Million in billing expense
• Saved $250,000/month in service order errors
• Reduced computer system outages by 74% in just six months
• Eliminated 90% of unnecessary repair appointments (8,000/month)

We got more from one day of Jay Arthur's Six Sigma consulting than we did with $120,000 from a big eight accounting firm.
    - Jim Mohrhauser, U S West

Six Sigma is a registered trade and service mark of Motorola, Inc.

Instructor Guide

Table of Contents

This Instructor Guide can be used by itself or with the Six Sigma Simplified Team Member Workbook

Six Sigma Simplified

FOCUS

IMPROVE

SUSTAIN

HONOR

Making Six Sigma Pay Off 3

Course Planning 12
   Show-Do-Know 14
   Course Agenda 15
   Course Roadmap 16
   Overview Module 18
   Laser Focus Module 20
   Problem Solving Module 22
   Process Management Module 26
   Short Course Roadmaps 29

Six Sigma Overview 32
   SCORE Model 33

Laser-Focused Improvement 35
   Voice of the Customer 37
   Identify the Indicators 40
   Develop a Master 6σ Story - Tree Diagram 44

Breakthrough Improvement 46
   Double Your Speed 46
   Value-Added Analysis 52
   Double Your Quality 54
   Step 1-Define the Problem 56
   Step 2-Analyze the Problem 60
   Step 3-Countermeasures 62
   Step 4-Results 64
   Cost of Poor Quality 71

Sustain The Improvement 72
   Define the Process-Flow Charts 74
   Sampling 80
   Check Stability and Capability 81
   Control Charts 87
   Monitoring System 97
   Benchmarking and Reengineering 98

Design for Six Sigma (DFSS) 99
   QFD 100
   Pugh Concept Selection Matrix 101
   Failure Modes and Effects Analysis (FMEA) 102
   Design of Experiments 104
   Block Diagram 105

Six Sigma Action Plan 106

Reading List 113

Other Tools 114
   Affinity Diagram 115
   Cost of Quality 116
   Force Field Analysis 117
   Histogram 119
   Matrix Diagram 120
   Systems Diagrams 123
   Tree Diagram 126

Instructor Guide

Six Sigma Simplified

This Instructor's Guide is designed to make learning the principles and processes of Six Sigma more easy. If you want to save $250,000 or more every time you do a project, then just follow the methodology in this book.

Robin Hood

1. If you don't know anything about Six Sigma or the improvement processes, read the 6σ Story in Six Sigma Simplified. Stories make learning fast, easy, and fun.

Six Sigma

2. If you don't know anything about Six Sigma, then consider starting on the next page with Making Six Sigma Pay Off. There are lots of ways to implement improvement projects; most are doomed to failure if you don't understand the science behind the implementation of change.

3. The next several pages offer lesson plans for training Six Sigma improvement teams.

Breakthrough Improvement

4. The rest of this book covers key tools and processes of how to:

- **Focus your improvements** with laser-like precision for maximum benefit at minimum cost
- **Improve key processes** to make them better, faster, and cheaper
- **Sustain the improvements** to ensure you don't slide back
- **Honor your progress** to recognize and reward the pioneering efforts of your employees.

The Core of Six Sigma

5. There are four key elements of Six Sigma:

- **Root-Cause Analysis** to reduce or eliminate *defects* in existing products or services.
- **Value-Added Analysis** to reduce or eliminate *delays* in existing products or services.
- **Process Management** to monitor and sustain the new levels of performance (e.g., lower defects-higher quality, less delay-faster) in existing products or services.
- **Design for Six Sigma (DFSS)** to create new products or services, and the processes for delivering them, that will start at 4.5 sigma (1,000 PPM) vs 2-sigma for most new products.

Instructor Guide

Making Six Sigma Pay Off

Six Sigma Targets

Sigma (σ)	Defects/Million
1	690,000
2	308,733
3	66,803
3.5	Average
4	6,210
5	233
6	3.4

In *Built To Last*, (Collins 1997), the authors mention the need for a BHAG or Big Hairy Audacious Goal. Using Six Sigma as a guide, you can measure your current performance in defects per million and set a BHAG of reaching the next level sigma.

So, if your computer system has 2% downtime, that's 20,000 minutes per million or about 3.5 sigma. Set a goal to reach 5 sigma (1 minute/5,000 minutes available)

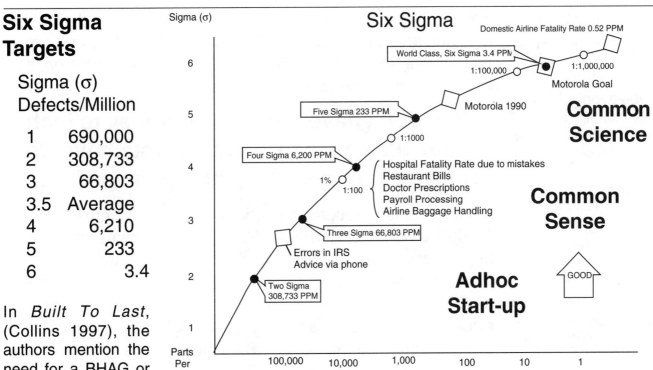

What Is Six Sigma?

What is Six Sigma? Six Sigma is a results-oriented, project-focused approach to quality. It's a way of measuring and setting targets for reductions in product or service defects that is directly connected to customer requirements. These reductions in the cost of poor quality translate into cost savings and competitive advantage. Sigma, σ, represents one standard deviation from the average or mean. Most control charts set their range at ±3s, but Six Sigma extends three more standard deviations. At six sigma, there are only 3.4 parts per million (PPM) defective.

Why Six Sigma? Why now? For the last few years, GE has been applying Six Sigma to improve GE's performance. GE invested $450 million to achieve $2 billion in savings by year end 1998 (Wall Street Journal, April). 1-2% of the employees are full-time "Black Belt" improvement leaders. At GE, 40% of executive bonuses will be based on achieving Six Sigma goals. Other Fortune 100 companies are following suit. Respected businesses embracing Six Sigma and customers demanding higher quality will drive the demand.

Making Six Sigma Pay Off

If we applied Six Sigma to ...

- Tax returns, there would only by 340 defects in the 100 million returns filed each year.

- Teen pregnancy—there would only be 34 pregnancies a year instead of 1,000,000.

- There would only be 3.4 accidents for every 1,000,000 miles driven.

Fortune 500 corporations invested millions in TQM during the late '80s and early '90s, with little to show for it. Unfortunately, the intuition to learn improvement methods was right on, but the implementation was often diluted

You don't have the time or money to inch your way to world-class; you have to make a number of "quantum leaps."

Most business processes produce about 2-3% errors or 20-30,000 defects per million. Relying on incremental (10%) improvements to get you to world-class quality would take an infinite period of time. Four Sigma is 6,210 defects per million—an 80% reduction. Five Sigma is only 233 defects/million—an additional 96% reduction. Six Sigma is 3.4 defects per million—an additional decrease of 98%. Continuous improvement won't get you to Six Sigma, but Breakthough Improvement will!

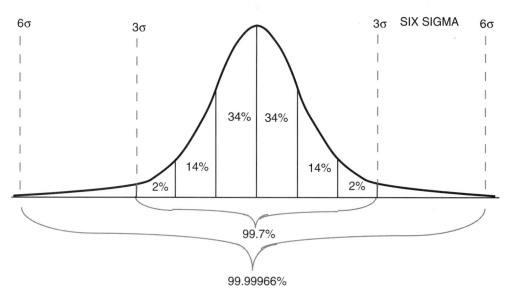

I'm not suggesting you throw away what you know or what you've learned so far; I'm suggesting an approach to <u>implementing</u> improvement efforts that you might want to consider if you want your investment in Six Sigma to payoff. What I'm about to offer will sound like heresy, but consider all of it before you make up your mind.

Instructor Guide

Making Six Sigma Pay Off

1. Forget The Quality Teams – Use SWAT Teams

Instead of letting teams choose their focus, consider 2-day leadership meetings to define and select key objectives.

SWAT Teams Instead of teams that meet indefinitely, consider having 2-day root cause "sessions" that bring together the right internal experts to focus on solving a critical business problem that affects customers and therefore profitability. These meetings focus on analyzing and verifying the root causes of problems and then identifying solutions. There are "instant" solutions that can be implemented immediately by the meeting participants, and there are "managed" solutions that need some leadership and project management to ensure proper implementation.

2. Forget Brainstorming – Use Existing Data

Brainstorming for problem ideas tends to be more like lighting a match in a dark room rather than using a laser pointer.

BHAGs First: Have operational leaders set a BHAG (Big Hairy Audacious Goal) for improvement. Using Six Sigma as a guide, where are you now? Set the next level Sigma as a target. If you're at 3-Sigma, go for 4-Sigma, and so on. Your target for world-class quality is "Six Sigma" or 3.4 defects per million widgets. If General Electric can save $2 Billion in one year by focusing on Six Sigma, what bonanza could you recover?

Next: Have operational managers <u>define</u> the problem to be solved using key performance measures–delay, defects or costs. If they can't develop a line graph and pareto chart that describes the "real" problem using data, then neither can anyone else.

3. Forget Classroom Training – Use JIT Training

2 Hour JIT Training Participants forget 90% of what they have learned in 48 hours unless they do something with it. Consider using 2-hour, just-in-time training for root cause meeting members the morning before they start solving a key business problem. This way, the training and experience will stick!

Instructor Guide

Making Six Sigma Pay Off

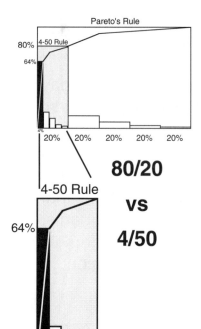

80/20 vs 4/50

4. Forget the 80/20 rule – Use the 4-50 Rule

When implementing most "new" ideas in an organization, leaders tend to treat the company like it is a 100-acre field. First they clear the whole field, then plow it, then scatter some seeds over it, and finally hope that a crop comes up. It should seem obvious, however, that companies of people are not exactly fields of corn. What if they were a different kind of field? What if you could get over 50% of the "harvest" from only 4% of the field? You can!

5. Skip the All-or-Nothing Approach to Implementation- Use the Crawl-Walk-Run Approach

Research into how groups of people adopt, adapt or reject changes proves that transformational change begins with less than five percent of the work force. The optimal way to introduce change into an organization is by treating the business like a field where four acres out of every 100 will produce over half the total harvest. This harvest will feed the ongoing transformation of the rest of the organization.

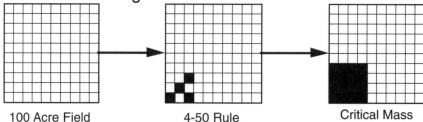

100 Acre Field → 4-50 Rule → Critical Mass

The initial 4%, the explorers, will convince the first colonists to cultivate even more land. This initial 4-16% constitutes the "critical mass" for a change to take root, grow, and spread rapidly across the organizational landscape. Trying to bring the whole corporate field under cultivation is reckless and invites wholesale rejection of the change.

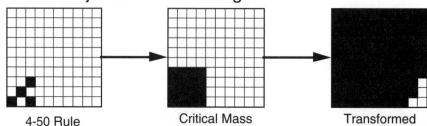

4-50 Rule → Critical Mass → Transformed

Instructor Guide

Making Six Sigma Pay Off

3 Sigma Process vs 6 Sigma Process

Remember that 80% of your effort only produces 20% of the harvest. Why bother? Rather than waste most of your hard-earned implementation dollars, focus and achieve results in just a few key areas. Then like seeds in the wind, word of mouth will spread the change to a few neighbors, and before long, virtually the entire company will have converted with minimal resistance and expense. It is never too late to shift your focus from whole field cultivation to laser-like cultivation. Just go back to basics and make a few key parts of the corporation successful.

6. Forget Inspection, Rework, and Scrap

A typical 3-sigma process has an inspection after key steps and a seemingly endless series of rework loops that clean up after the previous process, just like a parent with an unruly teenager. And all too often, the result has to be scrapped *and* the product or service has to be redesigned. Meanwhile, your customer is getting angrier and tired of waiting.

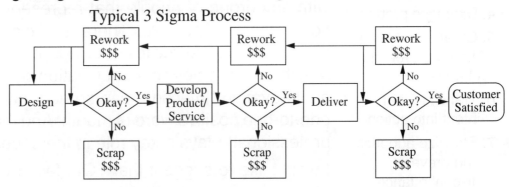

Typical 3 Sigma Process

But a 6-sigma process simply does not allow defects to occur. It prevents problems before they occur. The result is a better product or service, of higher quality, delivered faster that may have seemed possible.

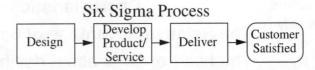

Six Sigma Process

Instructor Guide

Ensuring Successful Implementation

Change Agent Role

What is your role in getting people to adopt 6σ? How can you make it more contagious?

1. Prevent too much "adoption" (i.e., adoption by people who should reject it or delay implementation) which, paradoxically, will speed up diffusion.
2. Develop the need for change
3. Exchange information
4. Diagnose problems
5. Create an intent in the employees to change
6. Translate that intent into action
7. Stabilize adoption and prevent discontinuance
8. Achieve self-renewing behavior.

"The greatest response to change agent effort occurs when opinion leaders adopt, which usually occurs somewhere between 3 and 16 percent adoption in most systems."

For a technology as powerful as Six Sigma, it sure seems to be taking a long time to develop acceptance in the global village. The average period for the universal adoption of an innovation is 25 years. Question is: "Is there a way to speed it up? Is there a way to make 6σ more contagious in your company?" I believe the answer is "Yes!" So I'd like to offer for your consideration some information about how contagious ideas spread, what we can learn from it, and how to apply it to Six Sigma.

For over 50 years, researchers have studied (i.e., modeled) how changes are adopted, adapted or rejected by societies and cultures. This research is readily available in The Diffusion of Innovations, by Everett Rogers (Free Press, 1995). Diffusion is a model for understanding social change. There are several characteristics of "innovations" like 6σ that can be adjusted to increase the speed of adoption--advantages, compatibility, complexity, trialability, and observability. You might think of these characteristics as a way to develop rapport with any group of people that represent a culture--corporations, departments, etc. There is a clear decision strategy people follow to decide to adopt, adapt, or reject an innovation. And there are various communication channels through which an innovation "infection" can spread, although the winner is one-to-one positive word of mouth. And the change agent (6σ professional) plays a key role in the speed of adoption.

I would like to suggest that TQM failed to take root in some companies because the implementation failed to apply the lessons learned about diffusion--how changes are adopted by society. Let's use this change model as a filter for our experience. Along the way I'll suggest some possible ways to adjust our approach to increase the spread of 6σ.

Characteristics of Innovations

The Relative Advantage of 6σ

The heart of 6σ is about doing business better, faster, and cheaper. By letting customers get what they want, when they want it, at a price they perceive as offering superior value,

Instructor Guide

Accelerating Six Sigma

Adopter Categories

Innovators
(Explorers)–active seekers and champions of new ideas. They are intuitive and often perceived as deviant from the social system.

Early Adopters
(Colonists)–quickly notice shifts and begin to implement and improve them. Evangelists are essential for early adoption.

Early Majority
(Deliberate Settlers)–rely on the early adopters and opinion leaders to decide to adopt. These people need a coach.

Late Majority
(Skeptical Settlers)–These people won't adopt until it's safe. Neighbors from the early majority are important for later adopters.

Laggards
(Traditional) the corporate "immune system" will try to prevent a new idea from spreading.

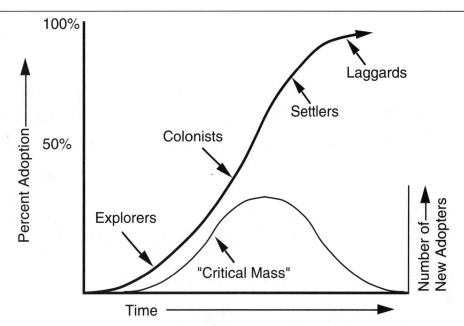

businesses thrive. 6σ, however, is often about preventing problems. Innovations involving prevention are slower to spread than innovations that solve pressing problems.

Compatibility - To be successful, any innovation must seek compatibility with a culture's:
- values and beliefs
- previously introduced ideas
- "felt" needs

Perhaps the most pervasive beliefs are that "we're working as hard as we can," as if working hard will solve the problem. This is a conscious "real world set of beliefs" that limit the adoption of 6σ.

Complexity - the degree to which an innovation is perceived as difficult to understand and use.

Because most people don't have a set of 6σ filters, they perceive the required math and graphics as complex and impenetrable. 6σ, for many people, invokes their limiting beliefs about their ability to learn, do math, or understand graphs. Howard Gardner's work on the seven intelligences suggests that only some of us are visual or mathematically oriented. To succeed, we must simplify and streamline 6σ and create a path for people to follow at their own rate of speed.

© 2003 Jay Arthur

Instructor Guide

Accelerating Six Sigma

Speed of Adoption

The **speed of adoption** can be affected by how it is done and who has the authority to make it happen. The adoption decisions (in order of speed of adoption):

Fastest: Optional–choices made by each individual (e.g., participation in 6σ)

Medium: Authority–made by one or a few people with power, status, or technical expertise.

Slowest: Collective–made by consensus of members

Varies: Combination of the above.

Trialability - new ideas, tried on the installment plan are easier to adopt because you can learn by doing.

People like to try things and then decide. Most employees want to make things better to serve customers more effectively. What else can we do to increase 6σ's trialability?

Observability - the ability to see results and for others to see them too.

This is a tough one, because the results of 6σ are often "invisible," because it prevents problems. There is often a delay between achieving better quality and increasing market share and profits. Recommendation: Find a way to make the effects of improvements "visible" to casual observers.

So those are the characteristics that can make 6σ more contagious–advantages, compatibility, complexity, trialability, and observability. Now let's look at the decision process used to adopt, adapt, or reject an innovation and the various types of adopters.

The Innovation-Decision Process

There is a step-by-step process people follow to decide to adopt, adapt, or reject a change or innovation in their lives:

1. First comes **knowledge** - an awareness/understanding of the problem. For me, this happened when the company decided to implement 6σ. This initial awareness always involves:

- What is 6σ?
- How-to apply 6σ (most important in trialing innovations)
- Why 6σ? -Principles underlying how it works
- Where might it apply in your life?

2. Next comes **persuasion**. I persuaded myself to learn 6σ. Other people need to be persuaded. How can we become the role models to influence more people to embrace 6σ?

3. Then comes the **decision**. Each person seeks information to decide whether to adopt or reject, either passively or actively, the change.

Instructor Guide

Accelerating Six Sigma

Communication Channels

A communication channel is a means by which messages get from one person to another.

Mass Media–TV, radio, and print–can:

1. Reach large audiences
2. Create and spread information quickly
3. Change weakly held attitudes

Interpersonal Channels

These are a slower but more effective means of persuading people to adopt a new idea. Used in the persuasion stage, it can:

1. Provide two-way exchange of information.
2. Allow an individual to form or change strongly held attitudes or beliefs

4. Then, assuming the person decided to adopt the change, they begin **implementation**. This sometimes requires adaptation of the change (e.g., applying 6σ in context of interest).

5. Finally, after a period of time, people achieve **confirmation**. They confirm for themselves that the change is a good one and that it deserves to stay. Or, they may decide to reject the change.

This decision process becomes increasingly complex when working with a system, group or organization rather than one individual.

Consequences of Adopting 6σ

- **Desirable vs undesirable**–6σ can help serve customers more effectively and grow market share, but it can also cause employee concerns. 6σ can also cost too much and deliver too little if we try to train everyone and implement too many teams.

- **Direct vs indirect**–the direct consequences of 6σ could be customer satisfaction (outcome) while indirect might be greater income (effect) from better personal and professional relationships.

- **Anticipated vs Unanticipated**–Will the gap between successful and unsuccessful companies grow as the successful use 6σ to develop greater marketshare? Will a focus on 6σ discourage innovation?

Recommendation: Amplify the benefits and prepare for potential problems. If we don't anticipate, understand, and resolve people's concerns about the undesirable, indirect, and unanticipated consequences of adopting 6σ, we really aren't concerned with the success of the change, are we?

Conclusions

We've covered over 50 years of research on how to accelerate the adoption and application of 6σ through:

1. better presentation of the perceived attributes of 6σ
2. understanding decisions and their effect on adoption
3. the power of mass media and personal communications

Instructor Guide

A Story For Instructors

My Story

Years ago, I spent five days in what turned out to be an excellent quality problem solving class. At the end, I knew that I had received some of the best training I'd ever had. I was so impressed that a few months later, I became an instructor. I went back to my job and started working with a quality team. <u>Two years later we gave up in disgust</u>.

During the same period, I became interested in neuro-linguistics--the science of how we run our neurology through language: pictures, sounds, and feelings. I took courses from one of the finest institutes in Boulder, Colorado. Each session would begin with one of the instructors telling a story: How the last remaining Japanese soldier was honored after 30 years in the jungle; how one instructor's grandmother took him fishing; and so on. I used to wonder: "What does this have to do with what we're here to learn?" I was so used to teachers talking at length about the theory and practice of things that this new style baffled me. After the story, the instructors would demonstrate the exercise that we were to do next. Then, we would go practice what we'd just seen demonstrated. I marvelled at how easy it was to learn what had originally seemed so foreign. And finally, the instructors would ask if there were any questions. Typically there were few. This pattern was repeated in each session, so I finally got used to it.

In the mean time, I had a chance to teach many of the five day quality classes. I watched my students as they worked on quality issues. Out of hundreds of teams, many of which lasted for 12 months or more; only a few were able to apply the quality improvement process and achieve useful results. Some of them spent six months just trying to get through step one! I also worked on several more successful improvement efforts; we were able to diagnose and resolve problems in anywhere from a few hours to a few days. I compared the teams that were successful with the teams that were not and figured out what made it possible for problem solving efforts to succeed <u>every time</u>.

The Power of Storytelling

After a couple of years of neuro-linguistics training, I was able to attend a trainer's training. "Finally!" I thought, "I get to learn the secrets of these great trainers. They began with a story from *Women Who Run With The Wolves*, called "La Loba." It was a story of an old medicine woman who lived in the desert and gathered up animal bones in an old hand sewn bag. She was especially fond of the bones of wolves. She would gather a complete skeleton, lay it out on the desert floor and begin to sing. As she sang, the wolf would begin to gather form until it stood up, looked at her and ran off into the night.

"What did this mean?" I wondered. I was soon to find out.

Instructor Guide

A Story For Instructors

Instructional Process

The instructors then introduced us to the **process** they had used all those years to teach us so efficiently. It was simple:

1. tell a mythical story that contains the pattern you want them to learn
2. tell a personal story that repeats the pattern you want them to learn
3. demonstrate how to do the pattern you want them to learn using examples from participants
4. have participants do the pattern you want them to learn (because they've now seen it at least three times.)
5. discuss any lingering questions

I started to use the neuro-linguistic method in the five day training. I scheduled two hours at the start of class to test the method. I told a story of how a team had identified and verified that cellular phones were causing false fire alarms. Then I asked the participants for existing problems that they were dealing with and I cast them in the form of the quality tools and processes. Using Six Sigma Simplified, I had them work on their own problems for an hour. I closely coached their exercise. Then I taught the rest of the course. Virtually all of the team leaders were able to form successful teams. Six Sigma Simplified and this manual are designed to optimize your effectiveness and efficiency as a 6σ Trainer by using the neuro-linguistics instruction process and Six Sigma Simplified.

When I looked at the five day training materials, I noticed that we spent a lot of time teaching tools that people rarely used. What's worse, some of these tools were so complex that they made people decide that quality was complex and difficult. Most people are phobic of math!

Now, in a two hour session, I teach only the quality strategies and the few key tools needed to achieve Laser-Focused Improvement results. And I do it using the neuro-linguistic process because it works! When I put Six Sigma Simplified in front of people, their faces light up with child-like enthusiasm. They think: "Maybe this quality stuff isn't so difficult after all." With this redesigned training process, people start using quality tools and strategies to focus, improve, or sustain their own processes in as little as *two hours*. Take a quantum leap in your instructional style!

Your Process

1. Six Sigma Simplified contains the metaphorical story. All you have to do is have participants read it!
2. Tell your own improvement story using data. People love stories.
3. Demonstrate the process and tools using examples from participants.
4. Have the participants do it. Coach them. Be the training wheels on their bicycle; they'll love you for it.

Using Six Sigma Simplified

How Adults Learn

Adults learn best from:
- Indirect experience (reading/hearing stories)
- Observing and imitating others (watching someone masterful)
- Direct experience (doing it themselves)
- Repetition

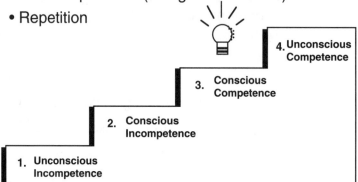

Show-Do-Know

Subtle changes in process can often lead to dramatic improvements in performance. To develop competence following the stairway above, most courses are taught following the Know-Show-Do strategy: Theory followed by demonstration followed by exercises unrelated to the job. It is possible, however, to develop unconscious competence directly! This guide follows the Show-Do-Know model: stories and demonstrations, followed by working on key issues, followed by Q&A to clear up any remaining confusion.

Unconscious Competence

Some useful ways of thinking about creating unconscious competence include:
- The mind can learn incredibly fast from patterns demonstrated repeatedly over a short period of time.
- There's always a simple, easy-to-understand way to do things
- **Three** is a charm—three repetitions, the three key tools of root cause teams are presented: line graph, pareto, and Ishikawa diagram, etc.
- Humans learn best when there are only 7 ± 2 pieces of information to learn. (There are only five steps in each process.)
- Participants need to work on their problems and processes to learn most quickly. Exercises must involve their jobs and real work.

Course Outlines

The following course outlines are designed to best use the adult learning strategies and Show-Do-Know to develop an inherent understanding of the underlying *process* of breakthrough improvement. Then, detailed understanding of the tools can be added.

Instructor Guide

Laser-Focused Improvement

Purpose		Acquire Essential Breakthrough Improvement Skills
Prework (1 hour before class)		• **Prework** - Read 6σ Story the night before class - Who are your key customers? What are their needs? - What is your core process to serve customers? - Identify a recurring problem to be solved - Bring data about the problem (defects, time, cost)
	60 min	1. **Overview--Focus, Improve, Stabilize, and Honor**
Improvement Planning (2 hours)		2. **Focus The Improvement Effort**
	30 min	- Tell participants an actual story of how the improvement process and tools were applied successfully.
	30 min	- Demonstrate the planning process.
	60 min	- Have them work in groups (if they share the same task) or individually to develop their own plan. Closely coach their efforts.
Problem Solving (4 hours)		3. **Improve The Process**
	45 min	- Tell participants an actual improvement story (e.g., fire alarm story)
	45 min	- Demonstrate the process using examples from the participants. Show them how their problem would fit into an improvement story.
	150 min	- Have them work in groups (if they share the same task) or individually to develop their own story. Closely coach their efforts.
Process Management (3.5 hours)		4. **Stabilize and Sustain The Improvement**
	45 min	- Tell participants an actual story of how stabilization improved the organization's performance.
	45 min	- Demonstrate stabilization (i.e., flowcharting and indicator selection) using examples from the participants. Show them how their process would fit into process management.
	120 min	- Have them work in groups (if they share the same task) or individually to develop their own story. Closely coach their efforts.
Limit		1.5 Days
Postwork (1-3 days)		• **Postwork** - Initiate planning, improvement, and stabilization. - Improve key problem areas. Stabilize the core processes.

Instructor Guide
Laser-Focused Improvement Road Map

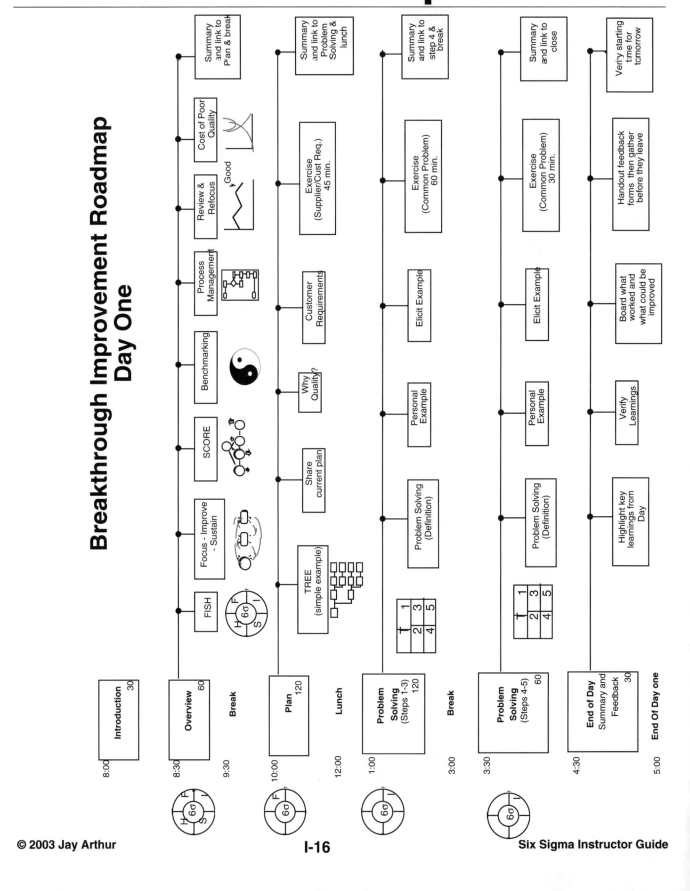

Instructor Guide
Laser-Focused Improvement Road Map

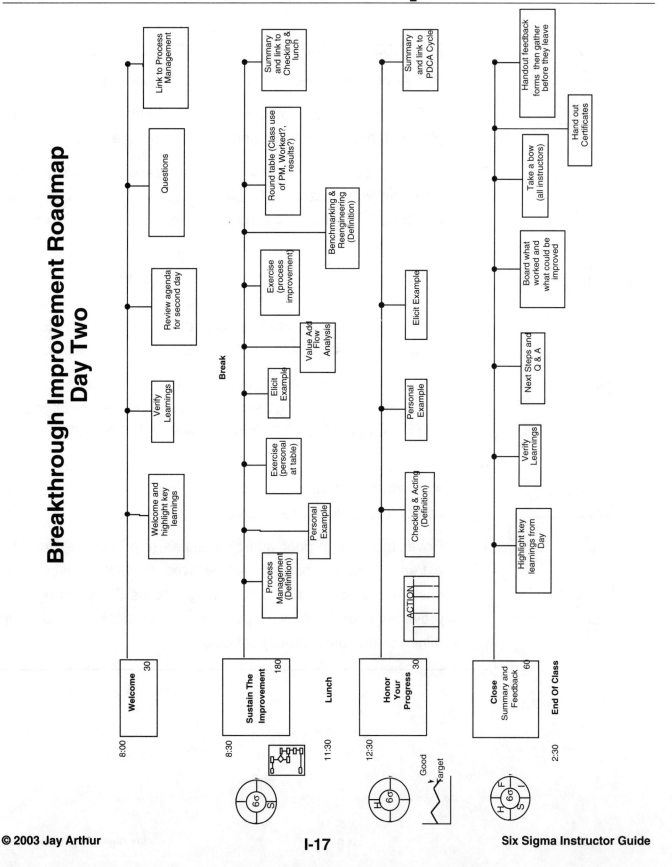

Instructor Guide

6σ - Overview

Step	Activity
1.	Begin with the FISH story on the next page. Relate this to the PDCA--plan, do, check, act cycle of continuous improvement.
2.	Define the three key processes of breakthrough improvement: planning, problem solving, and process management.
3.	Elicit from participants the current organizational vision and mission. This will be used for 6σ planning to focus all of the improvement efforts. Explain how 6σ planning serves to focus the improvement effort.
4.	Explain problem solving by using the SCORE model. Breakthrough improvement derives its power by moving from the current state (Cause->Symptom) to a desired state (Outcomes->Effects) by applying resources to key leverage points in the business.
5.	Explain how benchmarking and reengineering fit into continuous improvement by creating breakthrough improvements (2X to 10X) that then need continuous improvement and stabilization.
6.	Explain, briefly, how process management is the cornerstone of stabilization: defining and measuring processes to see if they conform to customer requirements.
7.	Cover the cost of poor quality. Failures (e.g., outages) can be extremely costly from the perspectives of waste, rework, and customer dissatisfaction. Preventing failures (breakthrough improvement) as well as inspection to detect failures can help minimize the total cost of poor quality.

Instructor Guide

6σ - Overview

FISH Story

```
Focus
Improve
Sustain
Honor
```

There is an old Asian saying:

Give a man a fish, and you feed him for a day

Teach a man to fish and you feed him for a lifetime.

To succeed at breakthrough improvement requires learning how to FISH:

- **Focus** the improvement effort on key problem areas to achieve the maximum benefit. (6σ planning)
- **Improve** the process through:
    - problem solving
    - value-added flow analysis
    - benchmarking
    - reengineering
- **Sustain** the improvement.
    - process management
- **Honor** your progress

Exercise

Purpose: To begin to evaluate the cost of poor quality

Agenda:

1. What are the costs of rework, and waste?
2. What contributes to inspection costs?
3. What does prevention cost? How is this course part of that cost?

Limit: 10 minutes in class (easel answers)

Instructor Guide

Focus

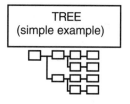

TREE
(simple example)

Customer Requirements

Exercise
(Supplier/Cust Req.)

Step	Activity
1.	Briefly explain why you think improvement planning is important.
2.	Tell an improvement planning story using a tree diagram and customer requirements matrix.
3.	Share the current improvement plan with participants.
4.	Explain how customers have requirements for products or services that fall into one of three categories: good, fast or value (cheap), which can be measured with either defects, time, or cost. Give a short example from the business.
5.	Have participants develop customer-supplier requirements for one key product or service, and expand the 6σ planning tree diagram to include their contribution.

Instructor Guide

Focus

6σ Planning Story

Using Post-it™ notes that you prepare before class, demonstrate the creation of a tree diagram using your organization's goals, objectives and measures. Use the customer requirements matrix to show the relationship between their requirements and the organization's measurements:

Tip: Use "real" business examples and stories. Simple, non-business related examples make it difficult for participants to bridge the learning into the real world. Or they will think improvement only works on simple stuff and check out.

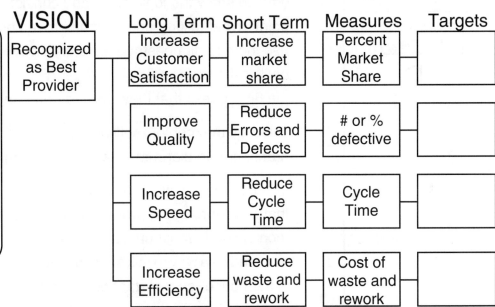

Exercise

Purpose: To begin developing the next level of objectives and measures.

Agenda:

1. Begin as a customer. Think of the main supplier. Identify one key requirement that you, as a customer, have in each area of good, fast, and cheap.

2. Now become the supplier to your customer. For your customers, identify one key requirement in each area of good, fast, and cheap.

2. Identify ways to measure these supplier-customer requirements with either defects, time, or cost.

3. Choose the <u>most important</u> of the three requirements and extend the development of the tree diagram to set long- and short-term objectives, measures, and targets for improvement.

Limit: 45 minutes

Tip: It is usually more challenging to have participants work on their own business issues, and it often delivers more value from the training time invested. As an instructor, you have to keep out of the content and keep them on track!

© 2003 Jay Arthur   I-21   Six Sigma Instructor Guide

Instructor Guide

Improve the Process
JIT Training - Steps 1-4

Problem Solving (Definition)

Personal Example

Elicit Example

Exercise (Common Problem) 60 min.

Summary and link to step 4 & break

Step	Activity
1.	Begin by telling an improvement story that, if possible, is related to the participant's business. Develop it on an easel, <u>one-step-at-a-time</u>, to show the logic, flow, and consistency of the story. At this point, avoid talking about the improvement process or tools, just use them to demonstrate the story's development. You will come back to the tools and process later. This will reinforce the pattern of the 6σ process that they learned from reading the 6σ Story.
2.	Describe the problem solving process and the three key tools--line graph, pareto diagram, and fishbone diagram.
3.	In class, discuss what they learned from the story. Everyone says that breakthrough improvement is just common sense, but common sense is most uncommon. What are the phrases they have acquired as a way to talk about quality (e.g., "clean the dirtiest room in the house" as a metaphor for working on the biggest issue).
4.	From the group, pull out one or two current issue(s) and use them to demonstrate and reinforce the first three steps again using their process or problem. Instead of perfect data, ask them for an approximation of current process performance; ask: "How could we measure that? What are the biggest contributors to the problem?" For example, if there are too many outages, you could demonstrate the problem solving process by showing: 1) a line graph of computer outages (number of minutes), 2) pareto chart of contributors--application software, system software, hardware, network, environment (e.g., power, air, etc.). If application software was the biggest contributor, you could show another level of stratification: Which applications caused the most outage? 3) fishbone diagram--corrupted or missing files. 4) countermeasures--prevention and backup

© 2003 Jay Arthur   I-22   Six Sigma Instructor Guide

Instructor Guide

Improve the Process

Problem Solving Story

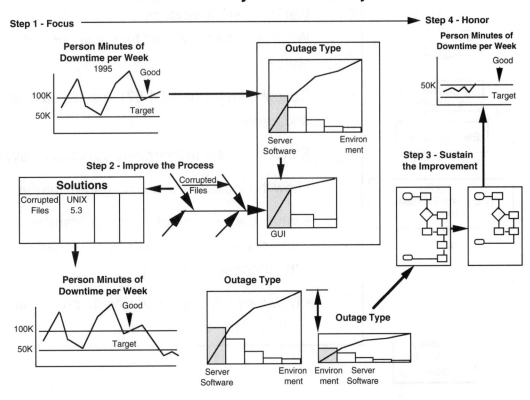

Distributed Systems 6σ Story

Exercise

Purpose: To apply problem-solving steps 1-3 (pages 55-63)

Agenda:

1. Identify one key problem area involving a customer requirement for good (defects), fast (cycle time), or value (cost of waste/rework).

2. Use this indicator to develop a line graph of current performance. If the team has historical data for the indicators, graphing the data can be done immediately. Otherwise, estimate and draw the graph.

3. Identify the most likely "Big Bar" on the pareto chart. Draw the pareto chart. Develop a problem statement based on the pareto chart.

4. Select the most likely main contributor to the problem and develop one bone of the cause-effect diagram to identify one root cause.

5. Develop potential countermeasures for this root cause.

Limit: 1 hour (15 minutes per step)

Tip: It is usually more challenging for participants to work on their own business problems, and it can deliver breakthrough improvements in class! As their guide, stay out of the content and keep them moving!

Instructor Guide

Improve the Process

Personal Example

Elicit Example

Exercise (Common Problem) 30 min.

Summary and link to close

Step	Activity
5.	Get feedback from steps 1-3 exercise. What worked? What didn't? What data will they need to gather back on the job to support their story?
6.	Describe step 4 (pgs 64-67): results, sustaining the improvement, and future plans.
7.	Either in class or in their exercise groups, have participants identify: • how the indicators should change if they've found the root causes • what they will need to do to standardize and replicate their improvements (e.g., develop flowcharts and ongoing tracking of the process indicators). • how they will know if they've improved enough, and what their future plans could include.

Instructor Guide

Improve the Process

Exercise

Purpose: To complete the 6σ Story

Agenda:

1. Identify how the indicators will change if countermeasures reduce or eliminate the root causes.
2. Identify ways to standardize, stabilize, and replicate the resulting improvement.
3. What next steps would the team recommend? Why?

Limit: 30 minutes

Instructor Guide

Sustain the Improvement

Step	Activity
1.	Tell a story of how process management was used to stabilize, sustain, and improve an existing process.
2.	Describe process management: flowcharting, indicators, and stabilization--root cause analysis and problem solving. Describe the main tools of process management: flowcharts, line graphs (control charts), and histograms.
3.	Select a process that all participants know or understand. Demonstrate the flowcharting process using Post-it™ notes on a sheet of easel paper. Walk participants through value-added flow analysis on this process: Which parts of the process are exception or rework loops (non-value added)? Which arrows have the most delay or idle time in them (non-value added)? Have participants do the flowcharting and value-added flow analysis exercise.
4.	Remind participants of customer requirements for good, fast, and value. Have them identify <u>one</u> "quality indicator" (e.g., outage minutes, missed commitments, or cost of waste/rework) based on customer's requirements. Have them identify one "process indicator" that would predict the quality indicator (e.g., software releases installed, delay, or file recovery).
5.	Briefly describe stability and capability. Refer them to an expert for help when they are ready.
6.	Exercise: Have participants develop indicators for their own process. If the team has historical data for the indicators, graphing the data can be done immediately.
7.	Give a brief overview of benchmarking (i.e., best practices) and business process reengineering. State how process management tools are essential to success in these efforts.

Personal Example

 Process Management (Definition)

 Elicit Example

Value Add Flow Analysis

Exercise (process improvement)

Round table (Class use of PM, Worked?, results?)

Benchmarking & Reengineering (Definition)

Summary and link to Checking & Acting

© 2003 Jay Arthur   Six Sigma Instructor Guide

Instructor Guide

Sustain the Improvement

Sustaining Performance Story	Using Post-it™ notes that you prepare before class, demonstrate the creation of a flow chart. Use the customer requirements matrix to show the relationship between the requirements and the quality and process indicators:
Exercise	**Purpose:** To begin developing a flowchart of a key business process **Agenda:** 1. For the participant's group, department, or organization, identify one key business process. Make sure they are focused before they start the exercise. 2. In sub-groups, develop a flowchart of the process. 3. Identify the non-value added rework loops and delays in the process. **Limit: 45 minutes**
Exercise	**Purpose:** To identify process and quality indicators for the process **Agenda:** 1. For the process previously flowcharted, identify one quality indicator based on the customer's requirements for good, fast, or value. 2. Identify up to two process indicators at key hand-off or decision points that will predict the process' ability to meet the stated customer requirement. 3. Based on the participant's knowledge, is the process stable? Capable? **Limit: 45 minutes**

Instructor Guide

Honor Your Progress

Personal Example

Elicit Example

Highlight key learnings

Next Steps and Q & A

Step	Activity
1.	Tell a story of how follow-up (checking and acting) was used to stabilize and improve an existing process.
2.	Describe the review and recognition phase of the overall improvement cycle.
3.	Elicit examples of existing mechanisms that help ensure breakthrough improvement.
4.	Highlight key learnings from entire course.
5.	Identify action plan to initiate breakthrough improvement efforts.
6.	Gather course feedback.

Instructor Guide

Breakthrough Improvement
Short Course Road Maps

Using Sections Of Six Sigma Simplified

- <u>Focus</u>, at a high level, identifies the vital few issues requiring transformation from present state to desired state (like a Compass pointing direction)
- <u>Improve</u> helps us move from the present state to the desired state.
- <u>Sustain</u> helps us define and stabilize the present state (Maps & measures)

2 Hour - Overview of any section (Focus, Improve, or Sustain)

1. Tell participants an actual story of how the improvement process and tools were applied successfully. (30 minutes)
2. Demonstrate using examples from the participants. Show them how their problem would fit into the improvement and stabilization process. (30 minutes)
3. Have them work in groups (if they share the same task) or individually to develop their own story. Closely coach their efforts. (60 minutes)

Laser-Focused Improvement Team (1-2 days)

- Prework
    - Identify problem to be solved
    - Gather data about the problem (defects, time, cost)
- Teamwork
    - Training (use 2 hour overview)
    - Solve the problem by facilitating the team through the process
- Postwork
    - Verify root causes, implement countermeasures, and measure results

Instructor Guide

Breakthrough Improvement
Short Course Road Maps

4-8 Hour - Improvement or Stabilization Overview

1. Tell participants an actual improvement or stabilization story from their company or industry.
2. Demonstrate using examples from the participants. Show them how their problems would fit into the improvement or stabilization story.
3. Have them work in groups (if they share the same task) or individually to develop their own story. Coach them closely.
4. Using example data, teach them how to develop line graphs, pareto charts, control charts, scatter diagrams, etc. <u>as they need it</u>.
5. Teach them more detail about the tools <u>as they need it</u> (e.g., the fishbone, Cause-Effect diagram).

Format of the Instructor Guide	Instructor Guide	Workbook
	The left-hand page is the Instructor Guide	The right-hand page is the page from Six Sigma Simplified.

Instructor Guide

Breakthrough Improvement
Overview

Objectives

- **F**ocus
- **I**mprove
  - Benchmarking
  - Reengineering
- **S**ustain
- **H**onor

In this section, you will gain a high level understanding of:
- The breakthrough improvement process.
- Focusing the improvement effort using planning
- Improving business processes using the SCORE model
- Sustaining performance using process management.
- Other improvement processes: benchmarking and reengineering.
- The costs of poor quality and how they cascade through an organization.

Key Processes

There are three key processes in breakthrough improvement:

- Focus- to laser-focus the improvement effort on a few key business priorities. This is also known as policy management or hoshin planning; hoshin means "shining needle pointing direction." Using the voice of the customer, business and employee, it is an ideal way to develop an annual "Master 6σ Story" that links and aligns all improvement efforts to achieve quantum performance improvement.

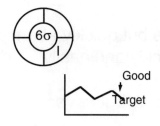

- Improve - to improve customer satisfaction by identifying and eliminating the root causes of problems involving time, defects, or cost. Also known as: quality improvement or root cause analysis; this process uses data to analyze problems and eliminate their root causes.

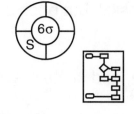

- Sustain - to define and stabilize any process. Also known as SPC--statistical process control, this process uses data to evaluate the ability of any business process to predictably and consistently meet the customer's requirements. It serves as a basis to systematically improve any process and maintain the gains from such improvements.

© 2003 Jay Arthur

Instructor Guide

Laser-Focused Improvement

Vision

Beginnings
"If we don't change our direction, we might end up where we're headed."   Chinese Proverb

> **VISION:**
> To be recognized as the best provider in the areas of:
> - Customer Service
> - Speed
> - Quality
> - Cost

Think of anything you've learned to do well. You learned to be a master by focusing on a specific aspect, improving in that area, and then stabilizing and sustaining the gains. If you've ever learned a sport like golf or tennis, you practice in one area at a time: serve, forehand, backhand, etc. (for tennis) or driving, irons, chipping, and putting (in golf). In music, first you master notes, then chords. Breakthrough improvement is a systematic path to mastery.

Mastery

In his book, *Mastery*, George Leonard describes the four types of learners he has encountered:

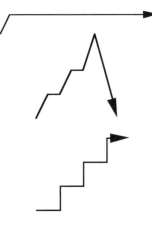

<u>Dabblers</u> - who keep trying different things but quickly abandon them if they don't produce results quickly and easily.

<u>Hackers</u> - who develop a certain level of skill, but never improve beyond basic skill.

<u>Compulsives</u> - who aggressively pursue a skill, but push so hard that they burn out.

<u>Masters</u> - who consistently focus, improve, and sustain ever higher levels of skill

Six Sigma Simplified

The Improvement Journey

The tide of evolution carries everything before it.
  George Santayana

In the long run, the only sustainable source of competitive advantage is your organization's ability to learn faster than its competition.
  Peter Senge

All improvement efforts follow a simple, basic process: FISH--Focus, Improve, Sustain, and Honor. FISH recognizes that there is a cycle of laser-focused improvement. Laser-focused improvement offers a systematic way to continuously improve every aspect of your business or personal life. Laser-focused improvement begins with **focusing** effort for maximum benefit, then **improving** the processes, **sustaining** the improvement and **honoring** your progress.

Breakthrough Improvement

FISH	Step	Activity
Focus	1	Focus the improvement effort
Improve	2	Reduce delay, defects, and costs
Sustain	3	Stabilize and sustain the improvement
Honor	4	Recognize, review and refocus efforts

FOCUS
Laser Focus

Most organizations suffer from diffused focus and conflicting objectives at all levels in the organization. To succeed with laser-focused improvement, organizations must shift from trying to do everything, to focusing on a few key issues that will move them the farthest toward satisfying customers.

Increase the Good

Decrease the Bad

All progress is based on the desire to move from a current level of success to a more desirable one--better quality, delivered faster and cheaper with a higher profit. Whether in business or your personal life, you will want to increase the good (profits and marketshare) and decrease the bad (waste or rework). To do both requires the application of resources. To move from the

© 2003 Jay Arthur   Six Sigma Instructor Guide

Instructor Guide

Laser-Focused Improvement

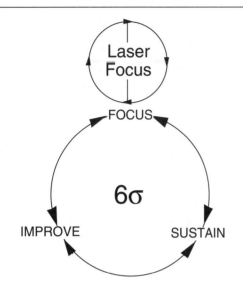

To achieve the maximum improvement, we must first focus the improvement efforts. "If I put you in a room with 50 rabbits, you will never catch all of them, but if I send them in a few at a time, you will catch all the rabbits."

We can't do everything so we must focus on the 20% that will give 80% of the return on investment. As little as 4% of your process accounts for 65% of the delay, defects, and waste. 6s planning is driven by the voice of the customer.

Then, we can begin to work in teams, or "learning organizations" to quote Peter Senge, to begin identifying and removing the "root causes" of delay, waste, and rework. Like a weed in a garden, we must pull up the root as well as the visible part of the plant (the symptom) to ensure that the problem will never come back.

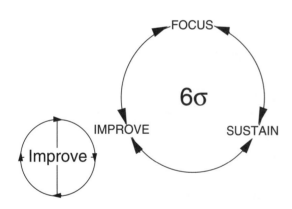

Improvement Overview
    It all boils down to:

Present State	Desired State
Poor	Rich
Low profit margin	Rewarding profit
Defects	Quality, reliability
Outage	Availability
Slow	Fast
Expensive	Value Leader

SCORE MODEL - Robert Dilts

Present State	Desired State
Cause->Symptom	Outcome->Effect

© 2003 Jay Arthur   Six Sigma Instructor Guide

Six Sigma Simplified
Focusing and Improving

current state of affairs to a higher, better state requires that leadership set the direction, and allocate the people, time, and money to make improvements happen.

Unfortunately, even in the smallest organizations, everyone has a different perception of what the desired state should be. To overcome this lack of focus, leadership uses quality planning to define the desired state--a vision of the desired outcomes and their effects.

The world owes all its onward impulses to men ill at ease. The happy man inevitable confines himself within ancient limits.
  Nathaniel Hawthorne

Then, leadership can define the few key improvement objectives, measurements, and targets that serve as a compass for moving from the present state to the desired state. By linking these objectives throughout the company, identifying improvement projects, and deploying the right resources we can begin the journey toward the company's vision. Leadership will also identify the core business processes and initiate improvement projects to define and dramatically improve each process.

IMPROVE Problem Solving

The problem-solving process further focuses improvements on the <u>root causes</u> of any quality problem--defects, high cost, or long cycle times. Preventing a root cause eliminates the symptoms, which ensures better products or services, increased customer satisfaction, and higher profits.

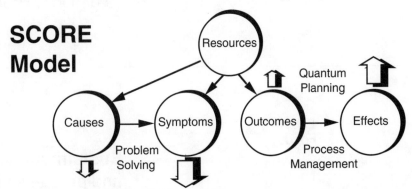

SCORE Model

A process that <u>can meet</u> the customer's needs at least some of the time will benefit from using the problem-solving process. A small reduction of root causes will significantly reduce symptoms and dramatically increase the positive effects of the improvement.

Instructor Guide

Laser-Focused Improvement

Playing Quantum Leap Frog

Benchmarking: Borrowing "with honor" outstanding processes from different parts of your own organization and from other companies.
Reengineering: Redesigning the process to deliver 10X improvements in speed, quality, and cost.

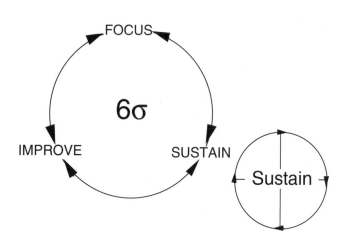

Once we have made improvements, we need to stabilize and sustain them. Have you ever stopped practicing a sport or a musical instrument for a few days and then come back to it? Process management sustains the improvement.

Sometimes, the process will need to be stabilized before it can be improved. In this case, process management will lead the improvement work (e.g., Core Processes)

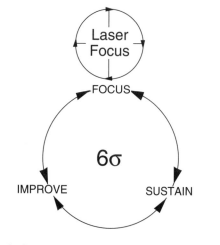

Sustain: Are you getting the results and progress you expect?

Honor: To reward success and redirect misguided improvement activities.

Six Sigma Simplified
Improving and Sustaining

IMPROVE
Value Analysis

Improvement teams can then identify the process steps that add value and those that don't. In most processes, only one out of every four steps add value. The rest either delay the product or service, or deal with the waste and rework found along the way. Focusing on the steps that add value and eliminating the steps that don't, often yields dramatic improvements.

IMPROVE
Design for Six Sigma

A process that is not even remotely capable of meeting the customer's expectations will need to be replaced. Any process over five years old may benefit from benchmarking (borrowing best practices) or reengineering to simplify and automate.

Plans get you into things but you got to work your way out.
    Will Rogers

Benchmarking compares your current performance with internal groups or external companies that are the best in class at any particular activity. Benchmarking seeks to understand the difference and adapt it to meet your internal business needs. It requires that you understand your own processes and performance first; so process management is a prerequisite.

Quality Function Deployment

QFD is a powerful tool for reengineering business processes or designing products and services. QFD identifies key customer needs and the essential people, process, and technology necessary to fulfill those requirements.

SUSTAIN
Monitor the Process

Improvement projects often begin by defining how work gets done--the present method of operation. Process management uses flowcharts and graphs to define the current process. Using the graphs, we can determine how "stable" or predictable the process is and how "capable" it is of meeting our customer's requirements.

Defined **Stable** **Capable**

HONOR

Throughout the improvement process, take time to celebrate, recognize, and reward improvement in both results and the application of all quality improvement processes.

Instructor Guide

Laser-Focused Improvement

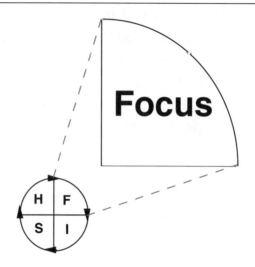

Introduction In this section, you will discover how to:

- Create a Master 6σ Story to focus and align the organization's mission to both the long- and short-term improvement objectives.

- Use the tree diagram, line graph, and VOC matrix.

- Select indicators to measure the customer's requirements and the progress of the improvement effort.

Key Tools There are three key tools in the focusing process:

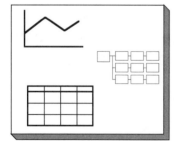

- Tree diagram - to focus and link the improvement objectives.

- Line graph - to measure customer requirements. This often provides step 1 of the problem solving process. (It is the basis for the control chart used to sustain the improvement.)

- Voice of the Customer Matrix - to analyze customer requirements and develop indicators.

Linkage The planning process feeds directly into doing:
- problem solving - to increase speed, quality, and cost by reducing cycle time, defects, waste, and rework.

© 2003 Jay Arthur   I-35   Six Sigma Instructor Guide

Laser Focus
Process

Focus

"Would you tell me please, which way I ought to go from here?"

That depends a good deal on where you want to get."

I don't much care where.

Then it doesn't matter which way you go."

-Lewis Carroll (Through the Looking Glass)

All quality improvements involve moving from a present way of satisfying customers to a more desired method. Before we can set the improvement processes in motion, however, we first have to define our direction for improvement. Planning for quality and initiating improvements follows the most basic of all quality processes, FISH--Focus, Improve, Sustain and Honor.

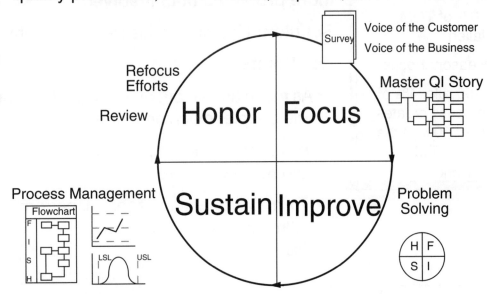

Process

Make no little plans: they have no magic to stir men's blood.
    Daniel H. Burnham

FISH	Step	Activity
Focus	1	Use the Voice of the Customer, business, and employee to identify desired long- and short-term objectives
	2	Identify and track the indicators
	3	Set targets for improvement
Improve	4	Initiate process improvements
Sustain	5	Sustain the improvements
Honor	6	Honor your progress
	7	Review and refocus objectives, teams, and improvement efforts as required

It is a bad plan that admits no modification.
    Publilius Syrus

© 2003 Jay Arthur — Six Sigma Instructor Guide

Instructor Guide

Laser-Focused Improvement
Focus The Improvement Effort

2X, 3X, & 10X Improvements

Breakthrough Improvement

- 50% of TQM failed
- Reason: Focus
- 4-50 rule: 4% of process creates over 50% of the waste and rework.

"The wider you spread it, the thinner it gets."
  Gerald Weinberg

Over 50% of TQM initiatives failed. The reason: failure to focus. The same is true of Six Sigma. Taking employees away from their work for days of training to solve trivial problems **creates more problems than it solves.**

Solution: Use data to laser-focus the improvement effort.

Key Points

- As much as $40 out of every $100 spent is for waste–scrap, idle people, machines, or materials–and rework.

- 4% of your business processes create over half of this waste and rework

- To make quantum leaps in performance from one level Sigma to the next, use data to focus improvements on this 4%.

Laser Focus

The Power of Focus

Each element of 6σ focuses attention on the key leverage points in the business--the "vital few." This is the essence of Pareto's rule: 20% of what you do yields 80% of the results.

Pareto's Rule

Process	Defects Time Cost
20 %	80%
4%	64%
1%	50%

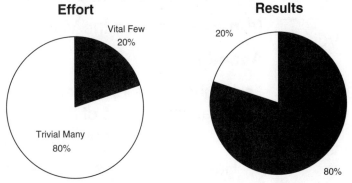

Pareto's Rule states that 20% of your effort delivers 80% of the results. 6σ seeks to identify the "vital few" areas that will deliver the most return on investment. 20% of your customers, for example, generate 80% of your revenue.

Pareto's rule can be applied to itself. 4% of your effort (20% of the 20% of the vital few) will yield 64% of the results (80% of the 80%). Using data, 6σ narrows your focus to create the maximum benefit for the minimum effort.

Six Sigma Improvement

- 50% of TQM fails
- Reason: Focus
- 4-50 rule: 4% of process creates over 50% of the waste and rework.

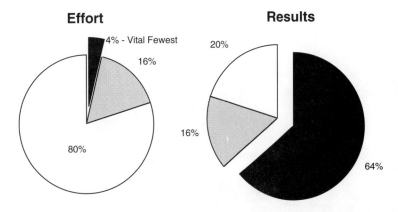

Even further, 1% of your effort will yield 50% of the result. For this reason, 50% improvements (a goal set with good old Yankee spirit) are often possible.

As you can see, you don't have to be perfect and fix everything, you just have to be able to focus on the vital few.

Instructor Guide

Laser Focus
Develop Voice of the Customer

Exercise

For improvement efforts to be successful, they must relate to the customer's requirements. The table of tables will tell you where to focus the improvement effort.

Purpose: VOC

Agenda:
- Voice
- Processes
- Relationships

Purpose: Develop the Voice of the Customer

Agenda:

- In one large or several small groups, have participants develop the voice of the customer (i.e., their requirements). Many common requests are shown on the table of tables.

- Have participants identify the processes and steps that participate in the delivery of the product or service.

- Have participants weight the relationship between the customer's requirements and the processes. Total the weights for each process. Which processes have the most impact on the customer's satisfaction?

Limit: 60-120 minutes

Identify Requirements
Voice of The Customer

Gather The Voice

The reason why we have two ears and only one mouth is that we may listen the more and talk the less.
  Zeno of Citium
  (c. 300 B.C.)

Purpose: to gather the customer's needs and wants as a basis for establishing objectives.

Only customers can create jobs. So customer satisfaction is the central theme of quality planning. There are <u>direct</u> customers (e.g., actual buyers or retailers) and <u>indirect</u> customers (e.g., government regulatory agencies). Each customer has unique requirements which can be related to the business.

Process

Step	Activity
1	Identify direct and indirect customers
2	Get the direct customer's requirements from surveys, focus groups, interviews, complaints, and correspondence. Review indirect customer requirements (e.g., regulations, laws, codes, etc.) Use the affinity diagram (pg. 42) to combine the direct and indirect customer requirements into key quality elements
3	Enter key customer voice statements on the left. Have customers rate the importance from 1 (low) to 5 (high).
4	Identify and enter key business functions for delivering the customers requirements along the top.
5	For each box in the center, rate the contribution of the "how" (top) to the "what" (left). Multiply the importance times the relationship weight to get the total weight. ● = Strong ○ = Medium △ = Low
6.	Total the columns. The highest scores show where to focus your improvement efforts.

Identify Requirements
Voice of the Customer

VOC

The Voice of the Customer uses <u>their</u> language to describe what they want from your business. Use a restaurant as an example to elicit the participant's "voice of the customer" for dining experiences.

When you go into a restaurant, what do you want?

Good	Get my order right
	I want good food
	I want an accurate bill
	Give me payment options--cash, check, credit card
Fast	Greet me and seat me promptly
	Serve me promptly
	Serve my food <u>when I want it</u> (fast or slow)
	Have my check ready
Cheap	Give me good value for the money spent
	Don't waste food

How do restaurants provide the meals? Greet & seat, ordering, preparing, serving, billing and collections. What are the most important processes?

Restaurant			Greet		Order		Prepare		Serve	Bill & Collect	
Relationships: ● 4 Strong, ○ 2 Medium, △ 1 Weak		Importance (1-5)	Greet	Seat	Take Drink Order	Take Food Order	Order Supplies	Prepare Order	Serve Order	Customer Check	Take Payment
Good	Get my order right	5			●	●		●	○		
	I want good food	5						●	○		
	I want an accurate bill	4			●	●				●	
	Give me payment options	3									●
Fast	Greet me and seat me promptly	4	●	●							
	Serve me promptly	5			●	●					
	Serve my food when I want it	5						●	○		
	Have my check ready	4								●	
Cheap	Give me good value for money spent	4					○	●	○		
	Don't waste food	3						●			
			4	4	12	12	6	16	8	8	4

Identify Requirements
Voice of the Customer

VOC Table — Using the form below, explore the interactions between the customer's requirements and your business.

Relationships
- ● 4 Strong
- ○ 2 Medium
- △ 1 Weak

Business Functions: Plan | Develop | Market | Deliver | Support

Importance (1-5)

Voice of the Customer (direct & indirect)

Customer Requirements:

Good
- Treat me like you want my business
- Give me products that meet my needs
- Products/services that work all the time
- Be accurate, right the first time

Fast
- If it breaks, fix it right the first time
- I want it **when** I want it
- Make commitments that meet my needs
- Meet your commitments
- I want fast easy access when I need help
- Don't waste my time

Cheap
- If it breaks, fix it fast
- Charge prices that are fair, competitive
- Help me save money

Total Weight

© 2003 Jay Arthur — Six Sigma Instructor Guide

Instructor Guide

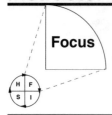

Laser Focus
Identify The Indicators

Customer Requirements

Customers only want three things from a product or service: It must be:

- **Better (i.e., good) -** high quality
- **Faster -** delivered or available *when they want it*
- **Cheaper -** perceived value for the price

Regardless of process, all critical user requirements fall into one of these three categories. They can usually be stated in simple, easy-to-understand, everyday language. These can then be easily measured by:

Requirement	Measure
Better	defects
Faster	delay
Cheaper	cost

Continuous Improvement Won't Get You To Six Sigma, But Laser-Focused Improvement Will!

Most business processes produce about 2-3% errors or 20-30,000 defects per million. Relying on incremental (10%) improvements to get you to world-class quality would take an infinite period of time. You don't have the time or money to inch your way to world-class, you have to make a number of "quantum leaps." Four Sigma is 6,210 defects per million–an 80% reduction. Five Sigma is only 233 defects/million–an additional 96% reduction. Six sigma is 3.4 defects per million–a further decrease of 98%.

Improvement Focus
Identify The Indicators

Purpose

Define specific ways to measure the customer's requirements and to predict the stability and capability of the process.

Indicators

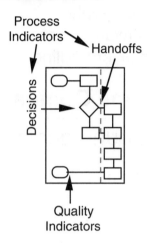

All problems invariably stem from failing to meet or exceed a customer's requirement. To begin to define the problem, you need to identify your customer's needs and a way to measure them *over time*--by hour, day, week, or month.

Quality indicators measure how well the product or service meets the customer's requirements. Process indicators, strategically positioned at critical hand off points in the process, provide an early warning system. For each "quality indicator" there should be one or more "in process" indicators that can predict whether you will deliver what your customer requires.

Requirement	Indicators	
	Quality or Process	**Period**
Better	Number of defects Percent defective (number of defects/total)	minute hour day
Faster	# or % of commitments missed time in minutes, hours, days	week month
Cheaper	cost per unit cost of waste or rework	shift batch

There are usually only a few key customer requirements for any product or service. Using the form on the next page, identify your main supplier, customer, the product or service used, and the process that creates it. Begin by identifying your requirements of the supplier. Then, identify your customer's requirements for the product or service. What do they want in terms of good, fast, and cheap. Then, based on your customer's needs, identify how you can measure it with defects, time, or cost. Finally, identify how often you will measure: by minute, hour, day, week, or month.

Goodness is uneventful. It does not flash, it glows.
- David Grayson

A stone thrown at the right time is better than gold given at the wrong time.
*   Persian Proverb*

Men go shopping just as men go out fishing or hunting, to see how large a fish may be caught with the smallest hook.
-Henry Ward Beecher

Improvement Focus
Identify The Indicators

Use examples from three different environments to demonstrate how to identify the indicators based on requirements. For a restaurant, software developer, or telephone company who are your:

Main customers? diners, application users, people
Main products/services? food & drink, software, connection
Main Processes? ordering, preparation, delivery, billing

Type	Restaurant	Software	Telephone Company
GOOD	Right Food Right Temp Fresh Friendly	Easy-to-use Bug free Accurate	Good sound quality Worldwide access
FAST	Prompt • seating • service • check	When I want it Timely Updates Fix it fast	Be responsive Available when I want it If it breaks, fix it fast
CHEAP	Value for $ Stop Waste	Value for $	Value for $ Help me be effective

Improvement Focus
Identify The Indicators

Exercise

Purpose:
Measures

Agenda:
- Supplier
- Customer

Limit: 30 minutes

Hint: It's often easier, as a customer, to first identify what you want from a supplier, then to identify what your customers want.

For improvement efforts to be successful, they must focus on the customer's requirements and ways to measure them-- defects, time, or cost.

Purpose: Develop the improvement measurements

Agenda:

- Identify one key supplier and one key customer.

- First, for one supplier, identify one requirement for good, fast, and cheap. Identify how you would measure the supplier's quality.

- Next, select one requirement for good, fast, and cheap from the Voice of the Customer. Identify an indicator for each of these requirements.

- Continue to expand the master 6σ story by adding the measures and any targets for improvement. If you're at three sigma, can you target four sigma? If you're at four, can you target five sigma?

Limit: 30 minutes

Instructor Guide

Laser Focus
Identify the Indicators

Exercise (Optional)

Purpose: Choose the improvement indicators

Agenda:

- In small groups, have participants define the customer's requirements and select indicators for good, fast, and cheap.
- How often will the indicator be measured?

Limit: 20 minutes

Purpose: Indicators

Agenda:
- Requirements
- Measurements

Improvement Focus
Identify The Indicators

Main: Supplier _____
       Customer _____
       Product or Service _____
       Process _____

Type	Requirement	Measurement	Period
Better		<u>Defects per million</u> (outages, inaccuracies, errors) <u>Defective per million</u> (scrap, rework, complaints) <u>Percent defective</u> (number defective/total)	
Faster		<u>Commitments missed per million</u> <u>Time to design, develop, deliver or repair or replace</u> <u>Wait or idle time</u>	
Cheaper		<u>Cost of rework or repair</u> <u>Cost of waste or scrap</u> <u>Cost per unit</u>	

Example

	Requirement	Measurement	Period
Better	What I want	Number of incorrect customer orders	Week
Faster	When I want it	Number of missed delivery commitments	Day
Value	Reduce my costs	Cost of correcting inaccurate orders	Day

Instructor Guide

Laser Focus
Develop Master 6σ Story

Key Tools There are three key tools in the focusing process:

 • <u>Tree diagram</u> - to <u>focus</u> and link the improvement objectives.

 • <u>Line graph</u> - to measure customer requirements

 • <u>Matrix diagram</u> - to analyze customer requirements and develop indicators

With these three tools you can focus the improvement efforts in ways that will create breakthrough improvements in quality and productivity.

Master Improvement Story Demonstrate the current improvement focus of the organization. Ask: What three key objectives will align in ways to achieve Laser-Focused Improvements in cycle time, defects, and costs?

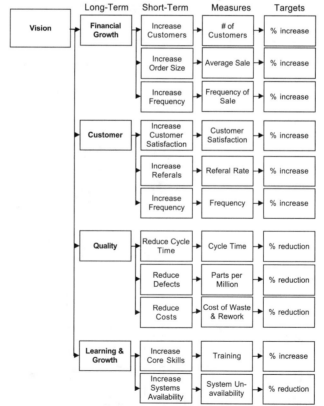

Exercise Use Post-it™ notes to develop a tree diagram of the key directions and improvement initiatives for your area.

Laser Focus
Create a Master Improvement Story

Master 6σ Story	**Master 6σ stories** link all of your efforts to ensure laser-focused improvements, not just incremental ones. The easiest way to depict a Master 6σ Story is with the "tree" diagram. Master 6σ Stories begin with a vision of the ideal world. This vision is then linked to long-term customer requirements, short term objectives, measures, and targets.
Key Tools 	**Why Is It So Important To Develop A Master 6σ Story?** 1) If leadership does it, they will commit to achieving it. 2) It links customer needs to the improvement efforts. This clear linkage, which is often missing, helps employees and leaders focus on the customer and align all of their actions to achieve customer outcomes, not internally generated ones. 3) Measurements based on customer requirements provide an ideal way to evaluate performance. 4) Detailed Master 6σ Stories can then be developed and linked to this one by individual managers. 5) Results can be measured and monitored easily.
Long Term Objectives	**Long Term Customer Requirements** invariably fall into one of three categories (from the voice of the customer matrix): • Better Quality--reliability and dependability • Faster Service--speed and on time delivery • Higher Perceived Value--lower cost
Short Term Objectives	**Short Term Objectives** translate these customer "fluffy" objectives into more concrete ones that can be **measured** and improved to meet the targets (from indicators): • Better Quality--fewer defects in delivered products, services, • Faster Service--reduced cycle time or missed commitments • Higher Perceived Value--greater benefits achieved by reducing the cost of waste and rework.
Targets	**Targets** are the BHAGs (Big Hairy Audacious Goals as James Collins calls them in *Built To Last*) that challenge our creativity and ability. 50% reductions in cycle time, defects, and costs are both challenging and achievable in a one year period. But to do so requires highly focused, not random, improvement work.

Instructor Guide

Laser Focus
Develop Master 6σ Story

Exercise

Purpose: Policy

Agenda:
- Long-term
- Short-term

Limit: 30 minutes

For improvement efforts to be successful, they must also link to the overall business objectives. The tree diagram helps create the "quality policy" which will tell you where to focus the improvement effort and what the improvement targets are.

Purpose: Develop the Improvement Objectives

Agenda:

- In one large or several small groups, have participants develop the long-term improvement objectives. Many common ones are shown on the diagram.

  Set a BHAG (Big Hairy Audacious Goal) for improvement. Using Six Sigma as a guide, where are you now? Set the next level Sigma as a target. If you're at 3-Sigma, go for 4-Sigma, and so on. Your target for world-class quality is "Six Sigma" or 3.4 defects per million widgets. If General Electric can save $2 Billion in one year by focusing on Six Sigma, what bonanza could you recover.

- Have participants identify annual improvement activities required.

- The next exercises will identify the measurements, targets for those measurements, and means of achieving them. Tell participants that they will continue to expand this tree diagram as we move forward.

Limit: 30 minutes

Laser Focus
Create a Master Improvement Story

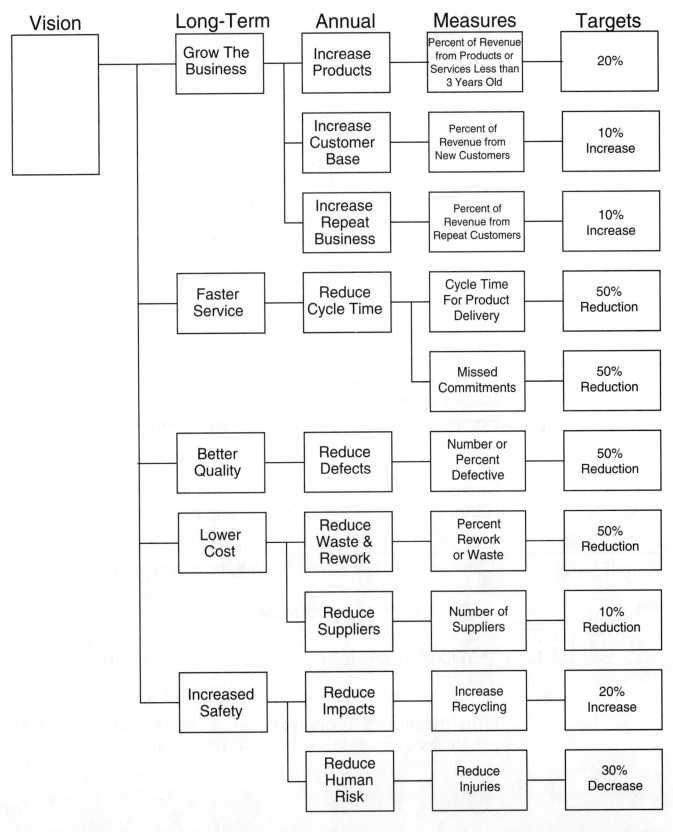

Instructor Guide

Breakthrough Improvement
Double Your Speed

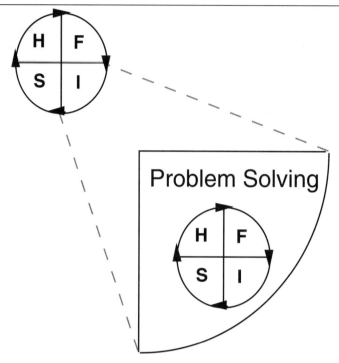

Introduction

Analyzing processes to eliminate delay and make them faster follows the FISH process:

Focus - to focus the improvement effort on key business processes
Improve - to reduce delay and rework
Sustain - to stabilize and sustain the improvements
Honor - to recognize, reward, and refocus efforts

Key Tools

There are <u>two key tools</u> in solving problems with speed:

- <u>Flow Chart</u> - to visualize the flow of work

- <u>Value Analysis Matrix</u> - to <u>identify</u> delays and rework

With these two tools you can eliminate 80-90% of all problems associated with speed.

Double Your Speed!

Value Analysis

There is always a best way of doing anything.
—Emerson

One of the best ways to improve your process is to find and eliminate as much of the delay as possible. Although the people are busy, the customer's order is idle up to 90% of the time—sitting in queue waiting for the next worker. Making existing processes faster follows the FISH process—Focus, Improve, Sustain, and Honor.

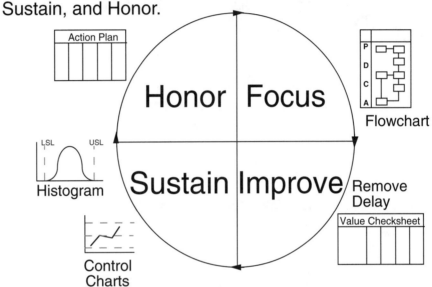

Process

FISH	Step	Activity
Focus	1	Define the process
	2	Identify the idle (i.e., wasted) time in the process (typically in the arrows)
Improve	3	Reduce or eliminate idle time
Sustain	4	Check the process for stability and capability
Honor	5	Recognize and reward improvement (Review and refocus improvement efforts)

© 2003 Jay Arthur   Six Sigma Instructor Guide

Instructor Guide

Breakthrough Improvement
Double Your Speed

Work in Progress

Employees often complain that they can't work any faster...and they're right...they can't, but the thing going through the process can work harder and faster. Any customer order lies idle a huge amount of time compared to how much time is actually spent processing it. That's why it is possible to dramatically reduce the time it takes to fill a customer's order without making anyone work one bit harder.

To understand how to reduce cycle time for products and services, you need to look at the spaces between the people or machines. That's where WIP, work in progress, stacks up. Find a way to reduce the volume between people or machines and dramatic improvements are possible. You may find bottlenecks in the middle of the process that can double your throughput.

Elicit Examples: Ask participants to name some examples of *delay* in their work or that they encounter when working with other companies.

Example: My wife's car was hit by a driver on a Friday morning. The driver didn't have their insurance ID card. By the afternoon (delay) when we got her number and could file a claim, it was too late to get the paperwork processed and a response from an adjuster. It was a holiday weekend (extra day's delay). Then it was two days before we heard from the adjuster late in the day (delay). Then it was too late to get the car to the repair shop for an estimate (delay) or get the paid rental car (delay).

It took seven days from the time of the accident to get work started on the car and a rental for my wife to drive to work. Now 3-4 weeks for repair.

Double Your Speed!

The employees are busy, but ...

If 4% of the business creates over 50% of the delay...

Why are you trying to speed up everything?

the order is idle 90% of the time

Reduce Delay!

Delay is caused by:
- <u>Idle time</u> between activities. Work in progress is up to 90% of delay.
- <u>Rework</u> to fix defects before, during, or after delivery.
- <u>Time wasted</u> creating products, services, or projects that end up being scrapped or cancelled.

Double Your Speed!
Define The Process

Eliminate Waste and Rework

How long does it take to build a 3-bedroom, two bath, two car garage home with all of the plumbing, fixtures, paint, carpet, and landscaped yard?

There is an annual contest to build house as fast as possible. **Last year's record was two hours and 48 minutes.** They do it by taking all of the idle time out of the process, combining steps, and getting all of the construction steps in the right order.

Value-added flow analysis assumes that <u>an idle resource is a wasted resource</u>. An activity or step that doesn't in some way directly benefit a customer is also wasteful. Rework, fixing stuff that's broke, is one of the more insidious forms of non-value added work: the customer wants you to fix it, but they really didn't want it to break in the first place.

Everyone is busy, but the order is idle:

- Requests for change may spend months in a prioritization cue before being worked (non-value added.)
- An order may sit idle waiting on an approval or material

On a process flow chart, **most of the non-value added time will be found in one of three places**:

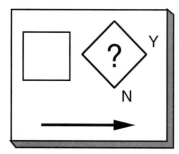

- **the arrows** (delay between processing steps)
- **rework loops** (fixing errors that should have been prevented)
- **scrap processes** (discarding or recycling defective products)

To eliminate these non-value added activities, how can you:

- combine job steps to prevent wasteful delay?
- initiate root cause teams to remove the source of the rework?

Double Your Speed!
Define The Process

Purpose Define the existing process as a starting point to begin improvement

Flowchart Symbols A flowchart uses a few simple symbols to show the flow of a process. The symbols are:

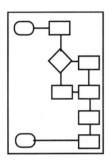

( )	Start/End	Customer initiated
Do It (box)	Activity	Adding value to the product or service (action verb & noun) (4% cause over 50% of defects)
◇ ? Y/N	Decision	Choosing among two or more alternatives (beware of rework loops)
→	Arrow	Showing the flow and transition (up to 90% of wasted/idle time)

Instead of writing directly on the flowchart, use small Post-it® notes for both the decisions and activities. This way, the process will remain easy to change until you have it clearly and totally defined. Limit the number of decisions and activities per page. Move detailed subprocesses onto additional pages.

Across the top of the flowchart list every person or department that helps deliver the product or service. Along the left-hand side, list the major steps in your process. In general, most processes have four main steps: planning, doing, checking, and acting to improve. Even going to the grocery store involves creating a list (plan), getting the groceries (do), checking the list, and acting to get any forgotten item. Virtually all effective business processes include these four steps.

The unified process of drawing and shooting was divided into sections: grasping the bow, nocking the arrow, raising the bow, drawing and remaining at the point of highest tension, loosing the shot.
 - Eugen Herrigel

Double Your Speed!
Define The Process

Process Flowcharts 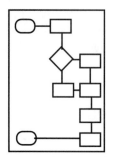	Process flowcharts extend the flowcharting technique to show "who does what" across the top of the flowchart and the macro steps of the process down the left-hand column. (See next page.) Guidelines for constructing process flow charts include: • Start with identifying customer needs and end with satisfying them. • Use the top row to separate the process into areas of responsibility • Use Post-it™ notes to lay out activities • Place activities under the appropriate area of responsibility.
Tips Flowcharting Quality indicators Process indicators	• Use square Post-it™ notes for activities and decision diamonds. • Draw arrows on any size Post-it™ note to show the flow, top to bottom, left to right. • Use smaller Post-it™ notes for process and quality indicators. • Participants will often offer activities at different levels of detail. As the higher level process flow gets more complex, keep moving subprocesses onto micro process diagrams. • Quality indicators which measure how well the process met the customer's requirements go at the <u>end</u> of the process. • Process indicators which predict how well the process will meet the requirements are most often placed at: 1) hand-offs between functional groups and 2) at decision points to measure the amount work flowing in each direction (this is most often useful for measuring the amount of rework required).
EXERCISE	**Purpose:** To develop a flowchart **Agenda:** 1. Closely coach participants to develop their own individual or team macro flowchart. If they have time, have them do the most important subprocess from the macro flow. **Limit:** 45 minutes

Double Your Speed!
Define The Process

Step \ Who	Customer				FLOWCHART
Plan	Request for product or service				
Do					
Check					
Act	Delivery of product or service				

Instructor Guide

Breakthrough Improvement
Double Your Speed

Removing Waste and Rework

Value Added Analysis (VAA) is the essence of reengineering--removing all of the wasted time and steps out of a process, simplifying, streamlining, and automating anything possible to error-proof and speed up the work.

Story

Working with a computer operations department, we used VAA to identify and eliminate 80-90% of the manual interventions by using the power of the job scheduler. This saved an hour of clock time per job, even more when techs were embroiled in a problem and didn't get around to checking the job for hours. It allowed the techs to spend their time productively fixing problems identified by the job scheduler.

Tips

- Avoid making the time spent too precise. Think yard stick, not micrometer; minutes, hours or days, not stopwatch.
- Rework loops are not value added. As Deming said: "We paid someone to put those defects into the product or service to begin with. Taking them out is waste."

Double Your Speed!
Refine The Process

Purpose   Identify the waste, rework, and delay that can be eliminated from the process.

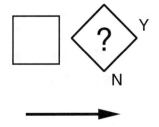

Over time, processes become cumbersome, inefficient, and ineffective. This complexity consumes more time and accomplishes less. Each activity, decision, and arrow on the flowchart represents time and effort. From the customer's point of view, little of this time and effort adds value, <u>most of it is non-value added</u>. From their point of view, delay and rework do not add value. We can increase productivity and quality by simplifying the overall process--eliminating delay and the need for rework.

Value Analysis

What we must decide is perhaps how we are valuable rather than how valuable we are.
-Edgar Z. Friedenberg

Everything is worth what its purchaser will pay for it.
- Publilius Syrus

What may be false in the science of facts may be true in the science of values.
- George Santayana

Step	Activity
1.	For each arrow, box, and diamond, list its function and the time spent (in minutes, hours, days) on the checklist.
2.	Now become the customer. Step into their shoes. As the customer, ask the following questions: • Is the order idle or delayed? • Is this inspection, testing, or checking necessary? • Does it change the product or service in a valuable way, or is this just "fix it," error correction work or waste?
3.	If the answer to any of these questions is "yes", then the step may be non-value added. If so, can we remove it from the process? Much of the "idle," non-value adding time in a process lies in the arrows: Orders sit in in-boxes or computers waiting to be processed, calls wait in queue for a representative to answer. How can we eliminate delay?
4.	How can activities and delay be eliminated, simplified, combined, or reorganized to provide a faster, higher quality flow through the process? Investigate hand-off points: how can you eliminate delays and prevent lost, changed, or misinterpreted information or work products at these points? If there are simple, elegant, or obvious ways to improve the process now, revise the flowchart to reflect those changes.

Instructor Guide

Breakthrough Improvement
Double Your Speed

EXERCISE

Purpose: To develop a value-added analysis

Agenda:

1. Using the process flowchart, have participants do a value-added flow analysis of the macro process. Where in their existing process is most of the wasted (idle) time and rework? What improvements could they initiate to eliminate the waste?

Limit: 45 minutes

	A	B	C	D	E
1		**Value Added Checklist**			
2	Activity, Decision, Arrow	Time Spent (hours, days, weeks, months)	Adds Value (not inspection or fix-it work)	Changes Product or Service Physically	Right The First Time (not waste or rework)
3	Claim Received	1 min	Y	Y	Y
4	Arrow to Establish Control	5 day	N	N	N
5	Establish Control	-	Y	Y	Y
6	Arrow to Create Folder	10 min	N	N	N
7	Create Folder	-	Y	Y	Y
8	Arrow to Need Evidence?	25 days	N	N	N
9	Need Evidence? (inspection/decision)	3 min	N	N	Y
10	No: Arrow to Exam Needed?	2 days	N	N	N
11	Yes: Arrow To Request Evidence	1 day	N	N	N
12	Request Evidence	45 min	Y	Y	Y
13	Arrow to Evidence Received? 90	60 days	N	N	N
14	Evidence Received? Days	-	N	N	Y
15	No: Arrow to Provide Letter	5-10 days	N	N	N
16	Yes: Arrow to Exam Needed?	2 days	N	N	N
17	Provide Letter	30 min	Y	Y	Y
18	Arrow to Exam Needed?	30 days	N	N	N
19	Exam Needed? (decision)	15 min	Y	Y	Y
20	No: Arrow to Rate Claim	-	N	N	N
21	Yes: Arrow to Request Exam	-	N	N	N
22	Request Exam	1 hour	Y	Y	Y
23	Arrow to Exam Complete? 36	-	N	N	N
24	Exam Complete? (decision) Days	25 days	N	N	Y
25	No: Arrow to Request Exam	-	N	N	N
26	Yes: Arrow to Rate Claim	90-100 days	N	N	N
27	Rate Claim	3 hours	Y	Y	Y
28	Arrow to Claims Review	1 day	N	N	N
29	Claims Review	2 days	N	N	Y
30	Arrow to Error Noted?	-	N	N	N
31	Error Noted? (Inspection/decision)	-	N	N	N
32	No: Arrow to Prepare Award	3-14 days	N	N	N
33	Yes: Arrow to Rate Claim (rework) 120	1 day	N	N	N
34	Prepare Award Days	15 min	Y	Y	Y
35	Arrow to Rating Correction Needed?	-	N	N	N
36	Rating Correction Needed? (Decision)	-	N	N	Y

© 2003 Jay Arthur

Double Your Speed!
Refine The Process

ACTIVITY, DECISION, or ARROW	TIME SPENT (hours, days, weeks, months)	NON-VALUE ADDED? (Y/N)		
		IDLE TIME, WAIT TIME, OR DELAY?	INSPECT? DETECT? TEST?	REWORK? WASTE? SCRAP?

Value Checklist

Instructor Guide

Breakthrough Improvement
Double Your Quality

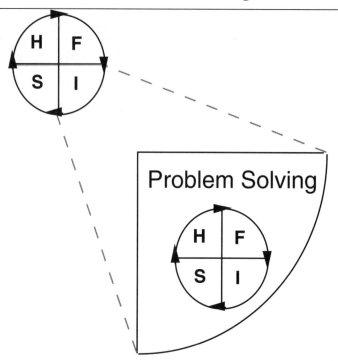

Introduction	Root-cause analysis is part of the overall improvement process and it follows the FISH process: **Focus** - to focus the improvement effort on key business problems **Improve** - to reduce delay, defects, and costs **Sustain** - to stabilize and sustain the improvements **Honor** - to recognize, reward, and refocus efforts

Key Tools	There are <u>three key tools</u> in solving problems with quality:

 • <u>Line graph</u> - to measure customer requirements

 • <u>Pareto chart</u> - to <u>focus</u> the root cause analysis

 • <u>Fishbone diagram</u> - to analyze the root causes of the problem or symptom

With these three tools you can solve 80-90% of all problems associated with defects, time, or cost.

© 2003 Jay Arthur

Improve Your Quality!

Problem Solving

Our problems are man-made, therefore they may be solved by man. No problem of human destiny is beyond human beings.
   -John F. Kennedy

Problems are only opportunities in work clothes.
   -Henry J. Kaiser

The Laser-Focused Improvement Process also follows the FISH process--Focus, Improve, Sustain, and Honor. It focuses on identifying problems, determining their root causes, and implementing countermeasures that will reduce or eliminate the waste, rework, and delay caused by these problems.

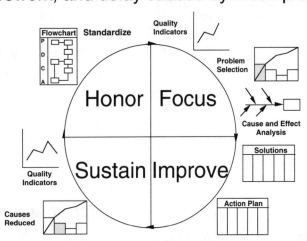

Process

The problem-solving process follows the FISH cycle to ensure continuous, never-ending improvement:

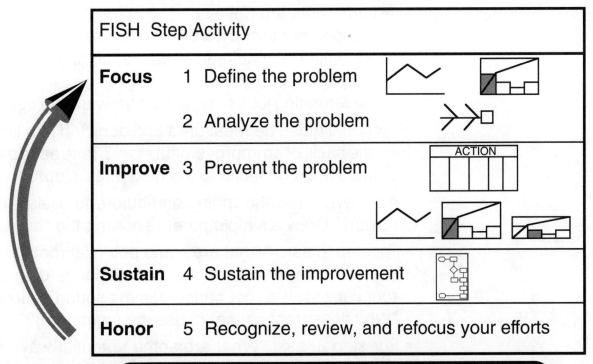

FISH	Step	Activity
Focus	1	Define the problem
	2	Analyze the problem
Improve	3	Prevent the problem
Sustain	4	Sustain the improvement
Honor	5	Recognize, review, and refocus your efforts

For more information, check out the Six Sigma Wizard at www.qimacros.com/qiwizard.html

Instructor Guide

Double Your Quality
Step 1 - Define the Problem

Step 1 Focus on Customer Satisfaction

There are only two types of problems: not enough of a <u>good</u> thing or too much of a <u>bad</u> thing, either of which should be measurable and easily depicted with a line graph. Since an increase in the "good" is often a result of decreasing the "bad," measures of the unwanted symptom make the best starting place.

Increase the Good

Since reducing the unwanted results of a process is often the best place to begin, the area of improvement can usually be stated as:

Decrease the Bad

Reduce the:
- **delay,**
- **defects,**
- **or cost**

in a product or service.

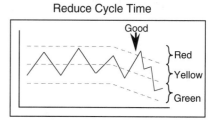

<u>Elicit from the participants</u> current problems in their work area. Identify these problems as delay, defects, or cost? Some examples include:

- complaints are defects
- outages or missed commitments are time problems
- the waste of media, floor space, computers, networks or people are cost problems
- the rerunning jobs or repair of hardware is a cost of rework.

How could these be measured and depicted in a line graph to form the basis of an improvement story? Take each participant's response and connect it to the next three steps.

- Ask: "What are the main contributors to (delay, defects, or costs)?" Draw a simple pareto showing the "big bar."

- For step 2, ask: "What are some potential root causes of the main contributor?" Draw a simple fishbone diagram of their root causes. If a root cause doesn't sound actionable, ask: "Why does (root cause) cause the (symptom)?"

- For step 3, ask: "What are some potential ways to prevent these root causes?"

© 2003 Jay Arthur   I-56   Six Sigma Instructor Guide

Laser-Focused Improvement
Step 1 - Define The Problem

Purpose Define a specific problem area and set a target for improvement

Problems are only opportunities in work clothes.
  -Henry J. Kaiser

There are two ways of looking at problems:
  Increase   (want more of a "good" thing)
  Decrease   (want less of a "bad" thing)

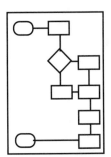

These are often two sides of the same coin:

an increase in ...	is equal to a decrease in . . .
quality	number or percent defective
speed	cycle time--to deliver a product or service
	idle time--people, materials, machines
profitability	cost of waste and rework

Measurement Solving problems is usually easiest when you focus on decreasing the "bad" rather than increasing the "good." Most problems can be easily expressed as a line graph showing the current trend and desired reduction in either cycle time, defects, or cost:

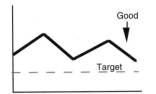

Example: Reduce defects in employee paychecks

Target: 15 or less

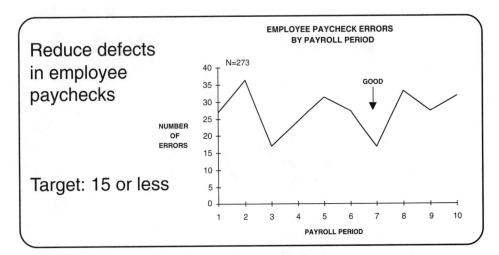

Instructor Guide

Double Your Quality
Step 1 - Define the Problem

Exercise

Purpose: Develop a line graph of the problem area

Agenda:

- In small groups, have participants select one indicator for defects, time, or cost. Using real or best guess data, have participants graph the current performance of the indicator.

Limit: 15 minutes

Purpose:
Problem

Agenda:
- Line Graph

Limit: 15 minutes

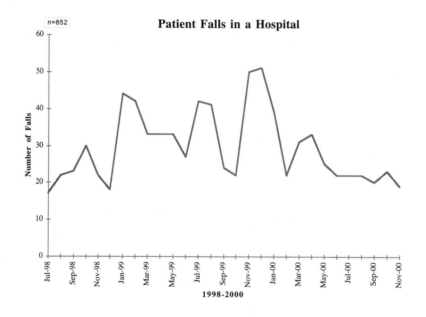

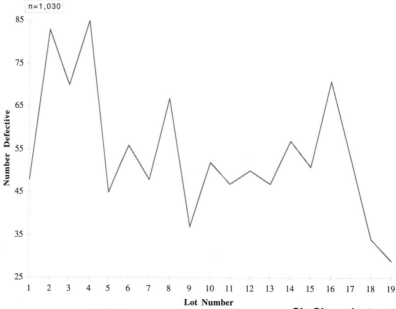

© 2003 Jay Arthur

I-57

Six Sigma Instructor Guide

Laser-Focused Improvement
Step 1 - Define The Problem

(circle one)

Problem: Reduce   Defects   in
                            Time   to deliver _____
                            Cost   to deliver (product or service)

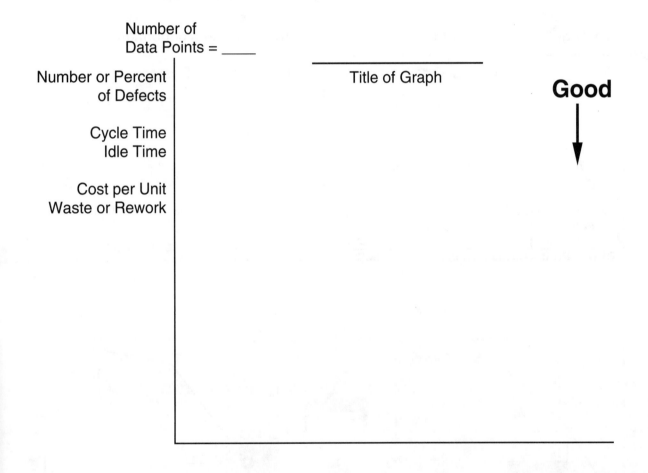

Who collected the data?

When was data collected?

Where?

What formula was used?

To automate all of your graphs, charts, and diagrams using Microsoft® Excel, get the *6σ Macros For Excel*

© 2003 Jay Arthur

Instructor Guide

Double Your Quality
Step 1 - Define the Problem

Narrow The Focus

To succeed at root cause analysis, you must first narrow the focus. Most improvement teams fail by trying to do too much. The pareto chart helps narrow your focus so that you can create dramatic improvements in the vital few problem areas. In the following example, teams could be working on CPU, print, and DASD costs, but we might still have a team focused on tape costs; specifically looking for ways to reduce waste.

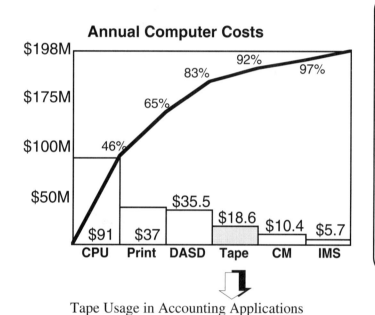

Definitions:
   use - cost *and* performance

Total Storage Costs:
   People (direct, support, svc center)

   Hardware

   Software

   Supplies (Tapes, labels, racks, etc.)

   Building (Floor space, power, off-site)

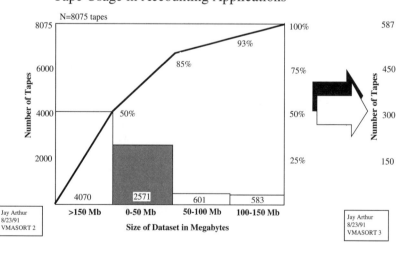

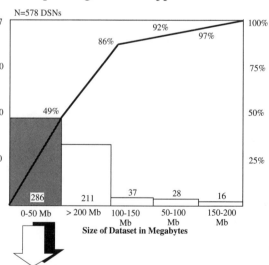

Problem Statement: During 1991, small tape datasets (wasting 3/4s of the tape available) accounted for 49% of all tape files, which was 10% of the total.

© 2003 Jay Arthur   I-58   Six Sigma Instructor Guide

Laser-Focused Improvement
Step 1 - Define The Problem

Pareto Chart

Problem *areas* are usually too big and complex to be solved with one effort, but when we whittle it down into small enough pieces, we can fix each one easily and effectively.

We only admit to minor faults to persuade ourselves that we have no major ones.
- La Rochefoucauld

This step uses the Pareto chart (a bar chart and a *cumulative* line graph) to identify the most important problem to improve first.

Often, two or more pareto charts are needed to get to a problem specific enough to analyze easily. The left axis shows the number of occurrences for each bar. The right axis shows the cumulative percentage for the line graph.

Begin by identifying the components of the problem:

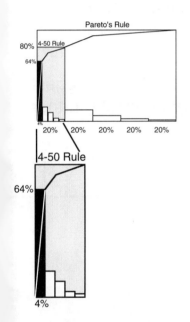

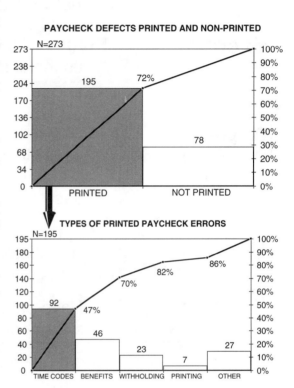

Indicator	Pareto Components
Defects	- types of defects
Time	- steps or delays in a process
Cost	- types of costs--rework, waste

A problem well stated is a problem half solved.

Once we have whittled the problem down to a small enough piece, we can then write a problem statement about the major contributor. This will serve as the basis for identifying root causes. We also need to set a target for improvement.

Problem Statement

Problem Statement: During the first five months of the year, time code errors accounted for 47% of all incorrect paychecks, which was 2X higher than the next highest contributor and resulted in 78 employee complaints.
Target: 50% reduction in time code errors

© 2003 Jay Arthur   Six Sigma Instructor Guide

Instructor Guide

Double Your Quality
Step 1 - Define the Problem

Narrow The Focus

Two or more pareto charts are often necessary to find a specific problem to solve. If the team doesn't narrow the focus here, they will end up with a "whalebone" diagram in step 2. Reinforce the link between the line graph and the pareto charts.

Exercise

Purpose: Develop a pareto chart of the problem area to laser-focus the analysis

Agenda:

- In small groups, have participants identify the main contributors to the problem indicator of defects, time, or cost. Using real or best guess data, have participants identify the biggest contributor to the problem (big bar on the pareto chart).

- If possible, have participants further stratify the biggest contributor or have them identify how they would further focus the problem.

- Have participants use the pareto chart to write a problem statement.

Limit: 30 minutes

Purpose: Define Problem

Agenda:
- Pareto Chart
- Problem statement

Limit: 30 minutes

Laser-Focused Improvement
Step 1 - Define The Problem

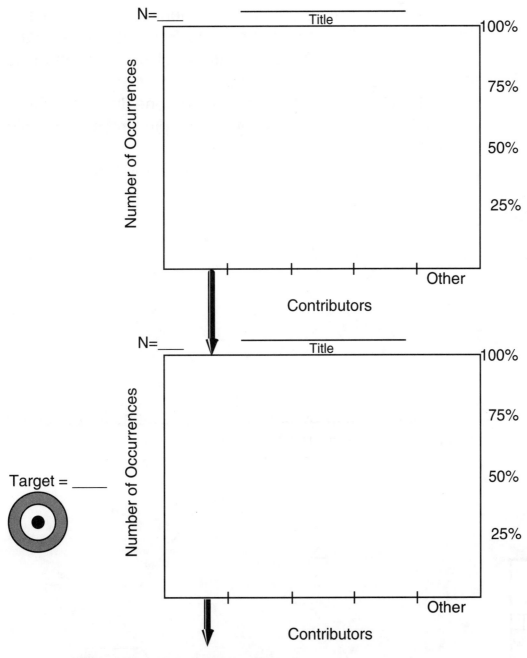

Problem Statement

During _____, ____, _____ accounted for ___% of _____,
       (Months) (Year) (Main Contributor)   (time, defects, cost)

which was ___ higher than desired and resulted in _____.
         (Gap)   (Pain)

© 2003 Jay Arthur   Six Sigma Instructor Guide

Double Your Quality
Step 2 - Analyze The Problem

Find The Root Causes

The Ishikawa, cause-effect, or fishbone diagram helps work backwards to diagnose root causes. For those unfamiliar with it, learning to use it can be frustrating, but, once learned, it helps prevent knee-jerk, symptom patching.

There are two main types of fishbone diagrams. One is a customized version of the generic--people, process, machines, materials, and measurement:

Customized Fishbone

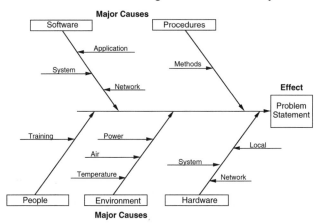

The other is a step-by-step, process fishbone that begins with the first step and works backward.

Process Fishbone

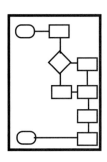

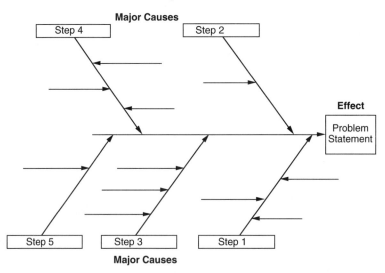

Remind participants to ask: "Why?" up to five times and to check their logic each time working up the chain saying "B *causes* A." Also remind them to verify their root causes before proceeding.

Laser-Focused Improvement
Step 2 - Analyze The Problem

Purpose

For every thousand hacking at the leaves of evil, there is one striking at the root.
—Thoreau

Identify and verify the root causes of the problem

Like weeds, all problems have various root causes. Remove the roots and, like magic, the weeds disappear.

Cause-Effect Analysis

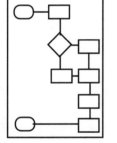

1. To identify root causes, use the fishbone or Ishikawa diagram. Put the problem statement from step 2 in the head of the fish and the major causes at the end of the major bones. Major causes include:

   • Processes, machines, materials, measurement, people, environment
   • Steps of a process (step1, step2, etc.)
   • Whatever makes sense

2. Begin with the most likely main cause.

3. For each cause, ask "Why?" up to five times.

4. Circle one-to-five <u>root</u> causes (end of "why" chain)

5. Verify the root causes with data (Pareto, Scatter)

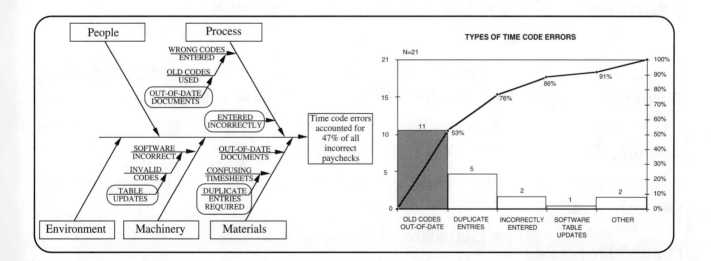

© 2003 Jay Arthur — Six Sigma Instructor Guide

Instructor Guide

Double Your Quality
Step 2 - Analyze The Problem

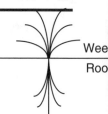

Find The Root Causes

The fishbone is not only useful for identifying the root cause of recurring problems (common cause variation), but it can also be extremely useful when stabilizing a process. Special causes of variation (e.g., power spikes, cable cuts, etc.) result in unstable processes as well.

Suppose there is a computer outage. The <u>problem statement</u> becomes: On Jan 31, 1996, System X went down putting 600 service reps out touch with the market unit tool set.

Major contributors to this problem can be identified and root causes determined. When collected over time, these special cause analyses will give you the data to cost justify the improvements necessary to prevent them.

Tar Pits

There are two main tar pits that teams fall into: whalebone diagrams and circular logic.

- **A whalebone diagram** (dozens or hundreds of bones) means that the problem wasn't focused enough in step 1. GO BACK and develop one more Pareto at a lower level of detail.

- **Circular logic** (C causes B causes A causes C again) invariably means the logic wasn't checked as it was developed.

Exercise

Purpose: Develop cause-effect diagram

Agenda:

- In small groups, have participants select the type of diagram and the most likely main contributor (big bone).

- Have participants ask "why?" up to five times to identify at least one root cause of the problem.

- Have participants discuss how they would verify this root cause using data.

Limit: 30 minutes

Purpose: Cause-effect

Agenda:
- Select Type
- Analyze root cause

Limit: 30 minutes

© 2003 Jay Arthur   Six Sigma Instructor Guide

Laser-Focused Improvement
Step 2 - Analyze The Problem

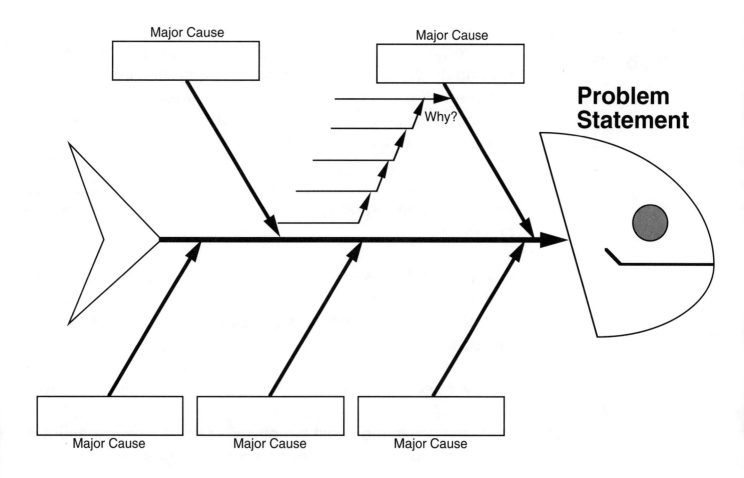

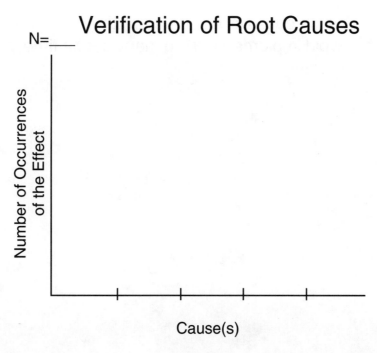

Verification of Root Causes

Double Your Quality
Step 3 - Countermeasures

Prevent The Problem

During the analysis of the problem, obvious countermeasures will often appear. The first three steps of the process tend to overlap. In the first few minutes of the first meeting, members will often offer unvalidated countermeasures and root causes. These should be captured and stored in the appropriate place in the improvement story.

Tip: Teams with unvalidated root causes can ruthlessly pursue worthless countermeasures, wasting time and money in the process.

Once the root causes have been validated with data to ensure that the team is tackling the true origin of the problem, various alternative countermeasures can be evaluated. There are two key questions:

- How <u>effective</u> is the countermeasure at preventing the root cause?
- How <u>feasible</u> (i.e., cost beneficial) is the countermeasure in terms of resources, time, and cost to implement.

Which countermeasures are the most effective and feasible? Avoid implementing too many at one time; one may cancel out another.

Laser-Focused Improvement
Step 3 - Prevent The Problem

Purpose   Identify the countermeasures required to reduce or eliminate the root causes

Take away the cause, and the effect ceases.
- Cervantes

Like ecological weed prevention, a countermeasure prevents problems from ever taking root in a process. A good countermeasure not only eliminates the root cause but also prevents other weeds from growing.

Defining Counter-measures

1. Transfer the problem statement and the root causes to the countermeasures matrix.
2. For each root cause, identify one to three broad countermeasures (what to do).
3. Rank the effectiveness of each countermeasure (Low, Medium, or High)
4. Identify the specific actions (how to do it) for implementing each countermeasure
5. Rank the feasibility (time, cost) of each specific action (Low, Medium, or High).
6. Decide which specific actions to implement.

© 2003 Jay Arthur   Six Sigma Instructor Guide

Instructor Guide

Double Your Quality
Step 3 - Prevent The Problem

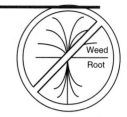

Prevent The Problem

The countermeasures diagram is a simple matrix used for prioritization. We just want to know how effective each countermeasure will be and the feasibility of each specific action required to put the countermeasure in place.

EXERCISE

Purpose: To apply steps 1-3 to business problems

Agenda:

1. Once you've lead 2-3 participants through the analysis of their root causes, have the whole class, either individually or in small groups, complete steps 1-3. Coach them closely to ensure that they understand the essence of the process. They can perfect the charts and graphs later. Identify one key problem area involving a customer requirement for good (defects), fast (cycle time), or value (cost of waste/rework).

2. Use this indicator to develop a line graph of current performance. If the team has historical data for the indicators, graphing the data can be done immediately. Otherwise, estimate and draw the graph.

3. Identify the most likely "Big Bar" on the pareto chart. Draw the pareto chart. While coaching participants through steps 1-3 have them focus their problem with two pareto charts and then develop a problem statement. The fill-in-the-blanks form provided will cover all of the aspects of a good problem statement. Develop a problem statement based on the pareto chart.

4. After participants have completed step one, coach them closely to put the problem statement in the head of the fish. Select the most likely main contributor to the problem and develop <u>one bone</u> of the cause-effect diagram to identify <u>one</u> root cause. Have them describe how they could verify the root causes with data.

5. Once the participants have completed steps 1-2, have them pretend that they have validated their root causes. Have them develop the countermeasures diagram for at least one root cause.

Limit: 2 hours

Laser-Focused Improvement
Step 3 - Prevent The Problem

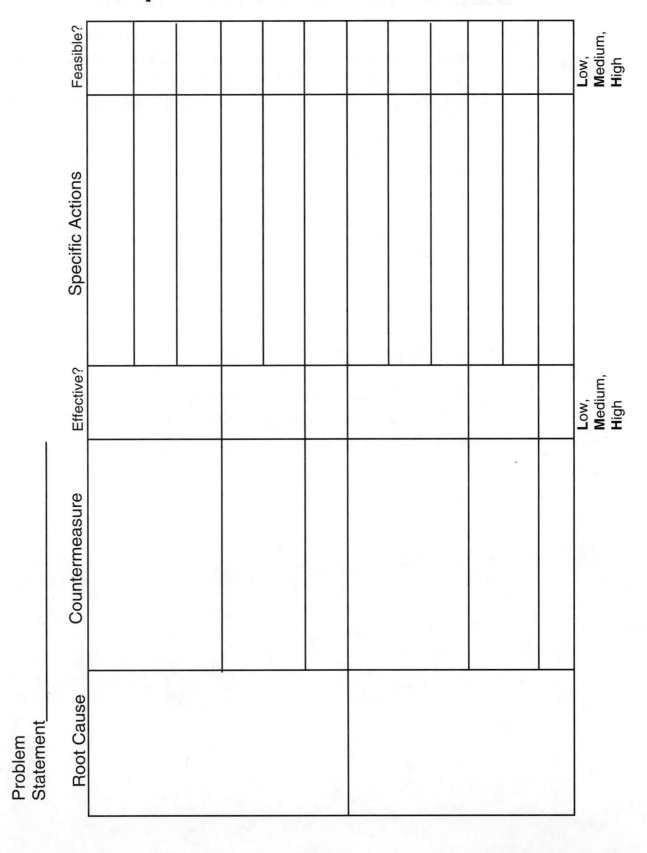

Instructor Guide

Double Your Quality
Step 3 - Prevent the Problem

Check The Results

There is only one way to determine if a countermeasure will actually reduce or eliminate the root cause--implement it and measure the effects. If it works, the symptom should decrease or vanish and the positive outcome and effects should increase.

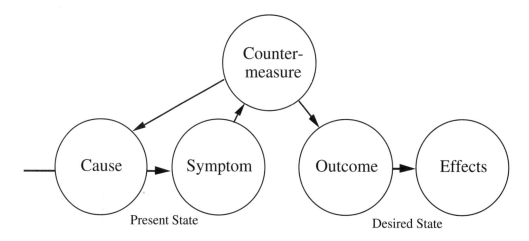

By using the line graphs and pareto charts developed in step 1, we can track the data after the countermeasure is implemented and show the change.

Laser-Focused Improvement
Step 3 - Prevent The Problem

Purpose Verify that the problem and its root causes have been reduced

Action should culminate in wisdom.
- Bhagavadgita

To ensure that the improvements take hold, we continue to monitor the measurements developed in step one and two. Both will improve if the countermeasures have been successful.

Verify Results

1. Verify that the indicators used in step one have decreased.

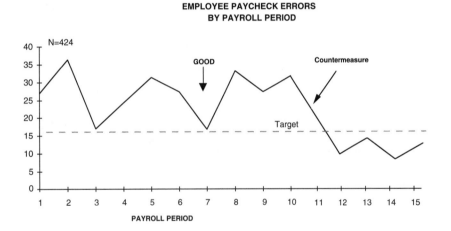

2. Verify that the major contributor identified in the Pareto chart in step one has been reduced.

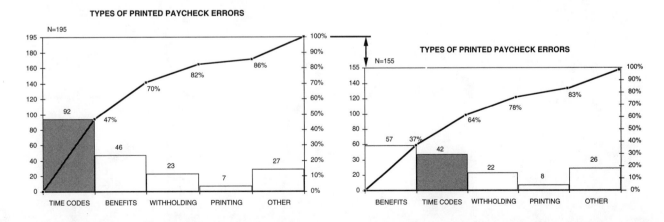

© 2003 Jay Arthur Six Sigma Instructor Guide

Double Your Quality
Step 3 - Prevent the Problem

Exercise (Optional)

> **Purpose:**
> Results
>
> **Agenda:**
> - Line graph
> - Pareto
>
> **Limit:** 20 minutes

Purpose: Graph the results

Agenda:

- In small groups, have participants use real or best guess data to anticipate the future performance of the indicators developed in step 1.

Limit: 20 minutes

Laser-Focused Improvement
Step 3 - Prevent The Problem

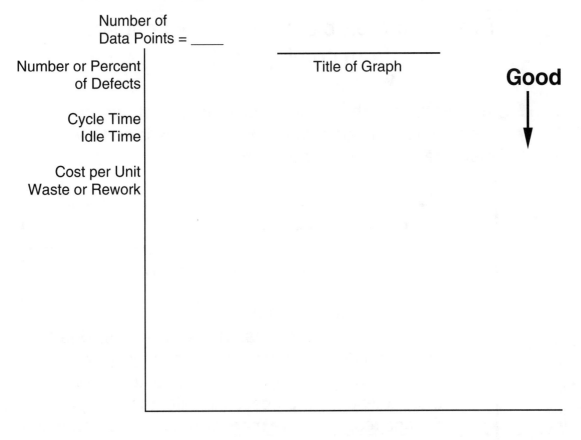

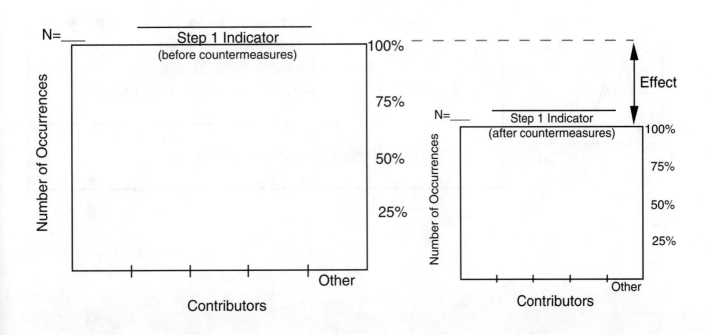

Instructor Guide

Double Your Quality
Step 4 - Sustain The Improvement

Purpose   Prevent the problem and its root causes from coming back

There is always a best way of doing everything.
-Emerson

To ensure that the improvements take root, we need to develop a flow chart of the improved process and a way to measure its ability to meet customer needs.

Develop Process System

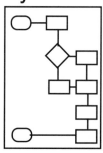

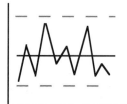

1. Flowchart the revised process.

2. Identify the key "process" and "quality" measures.

   Quality indicators (step 1 indicator) measure the customer's requirements. Strategically positioned at critical hand-off points, process indicators provide an early warning system. For each output indicator, there will be an in-process indicator that can predict the end result: Will we meet the customer's common requirements for:

Requirement	Measure
Good	Number of defects per step
Fast	Cycle time per step
	Missed milestones
Cheap	Amount of waste and rework

3. Chart and track these indicators

© 2003 Jay Arthur   Six Sigma Instructor Guide

Laser-Focused Improvement
Step 4 - Sustain The Improvement

Purpose	Prevent the problem and its root causes from coming back

None will improve your lot, If you yourselves do not.
- Bertolt Brecht

Like crops in a garden, most improvements will require a careful plan to ensure they take root and flourish in other gardens. To transplant these new improvements into other gardens will require a stabilization plan.

Lock in the Gains

ACTION				

Action Plan

WHAT? (Changes)	HOW? (Action)	WHO?	WHEN? Start / Complete	MEASURE? (Results)
People	Training			
Process	Define system and measures / Implement / Monitor			
Machines (Computers, vehicles, etc.)				
Materials (Forms & Supplies)				
Environment				
Replicate	Identify areas for replication / Initiate replication			

© 2003 Jay Arthur

Instructor Guide

Problem Solving
Step 4 - Stabilize

Action Plan   Once the process is stable and capable of meeting customer's requirements, the next most important step is to multiply the benefits of the improvement process. Breakthrough improvements are possible when teams share and apply their learnings.

Exercise

> **Purpose:** Sustain and Replicate
>
> **Agenda:**
>
> - Action Plan

Purpose: To complete 6σ Story

Agenda:

1. Identify how the indicators will change if countermeasures reduce or eliminate the root causes.

   (Since participants can't usually implement the countermeasures during training, we can ask them what to expect from this step. Assuming that the countermeasures where successful, what effect would the participants expect to see in the graphs from step 1 and 2?)

2. Identify ways to standardize and stabilize the resulting improvement.

   (Assuming that the countermeasures where successful, what steps would the participants expect to take in order to stabilize the process and lock in the improvements? What changes are necessary in people, process, machines, materials, or the environment? This can be handled as a discussion in class.)

3. Assuming that the countermeasures were successful, ask the participants: "Who else could benefit from what you've learned? How can this improvement be either <u>adapted</u> or <u>adopted</u> by other members of your organization to maximize the resulting benefit?

4. What next steps would the team recommend? Why?

Limit: 45 minutes

Laser-Focused Improvement
Step 4 - Multiply The Gains

Purpose — To increase the return on investment from each improvement effort.

Nothing succeeds like success.
  Alexandre Dumas

To ensure that we get the maximum benefit from having solved this problem, we have to get this improvement into the hands of all the other people who could use it.

Identify Places to Increase the Gains

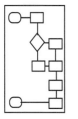

1. Brainstorm a list of potential clients for the improvement. Any group or individual that shares a similar customer, product, or service that could benefit from the improvement.
2. Select the key groups for replication.
3. Forward the process (flowchart and indicators) to the groups targeted for replication.
4. Follow up to ensure that the improvements have been implemented. Replicate any additional enhancements that have been made by these other groups.

WHERE? Where will this process be useful?	WHAT? What needs to be done to initiate?	HOW? How will the process be replicated?	WHO? Who owns the replication?	WHEN? Start / Complete	
		Adopt process Adapt process to fit Incorporate existing improvements			

Instructor Guide

The 6σ Story
Mapping the Journey to Six Sigma

Exercise (Optional)

Purpose:
6σ Story

Agenda:
- Focus
- Improve
- Sustain
- Honor

Purpose: Develop a short 6σ Story

Agenda:

- In small groups, have participants use everything they've developed until now to develop a complete 6σ Story.

Limit: 60 minutes

The 6σ Story

Mapping the Journey To Six Sigma

One Baby Bell reduced computer downtime by 74% in just six months using Laser-Focused Improvement. The next two pages show the actual improvement story.

Define The Problem

At the beginning, there were 100,000 "seat" minutes of outage per week. Since there were 9,000 service representatives, that means only 11 minutes of outage per week per person, but all totalled, it meant the loss of 1667 hours, 208 person days, or five person weeks. In other words, it was the equivalent of having five more service reps unavailable.

Target:

The VP of Operations set a goal of reducing this by 50% which caused a lot of grumbling, but on analysis, they found that 39% of the downtime was caused by the server software, 28% was caused by application software, and 27% server hardware.

Analyze the Problem

Multiple improvement teams tackled each of these areas. Root cause analysis and verification determined that password file corruption, faulty hardware boards, processes, and one application accounted for most of the failures.

Prevent The Problem

Multiple countermeasures were implemented including upgrades to the operating system in over 600 servers to prevent password file corruption and other problems.

Check Results

In less than six months they had exceeded the goal by achieving a 74% reduction.

Sustain The Improvements

A system was implemented to manage outages for both immediate and long-term improvement.

Instructor Guide

Six Sigma Improvement Story
Breakthrough Improvement Made Easy

Date: ___/___/___

DEFINE THE PROBLEM (Why do I think I have a problem?)

Laser-Focus The Improvement

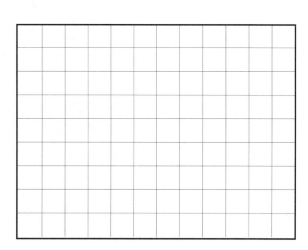

ANALYZE THE PROBLEM (What are the causes of the problem?) Cause and Effect

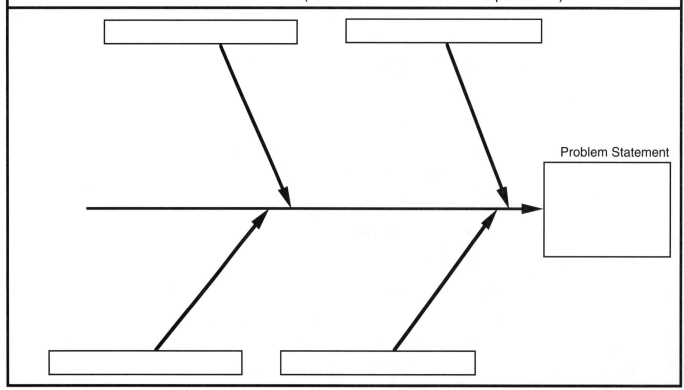

Problem Statement

© 2003 Jay Arthur   Six Sigma Instructor Guide

THE SIX SIGMA STORY
Problem-Solving Made Easy

Date: ___/___/___

DEFINE THE PROBLEM (Why do I think I have a problem?)

Reduce Minutes of Computer Outage

100,000 — Good — Target = 50%

J F M A M J J A S O N D
1995

Contributors To Computer Outage

100,000 — 100%
39k — 39% — 67% — 94%

Server Software | Application Software | IWS | DMP | Server Hardware | Network | Environment

Teams focused in all five areas to drive 50% reductions in outages

ANALYZE THE PROBLEM (What are the causes of the problem?) — Cause and Effect

Server Hardware
- Boards
- Hard Disk
- Power Supply

Server Software
- Password File Corruption
- Old UNIX version
- Process Errors

Environment — Power Outage

Network — Modem, Connection

Application Software
- Missing Files
- File Corruption
- IWS Software

Problem Statement

From Oct to Dec. 1994, server availability averaged 100,425 minutes of downtime per week.

Pswd Files | Board | Process | DMP | Pwr Supply

© 2003 Jay Arthur — 69 — Six Sigma Instructor Guide

Instructor Guide

PREVENT THE PROBLEM (What steps will correct the problem?)

Root Causes	Countermeasures	Effectiveness (H-M-L)	Specific Actions	Feasibility (H-M-L)

CHECK RESULTS (Is the suggested solution working?)

SUSTAIN THE IMPROVEMENTS (How do we sustain and replicate it?)

© 2003 Jay Arthur   Six Sigma Instructor Guide

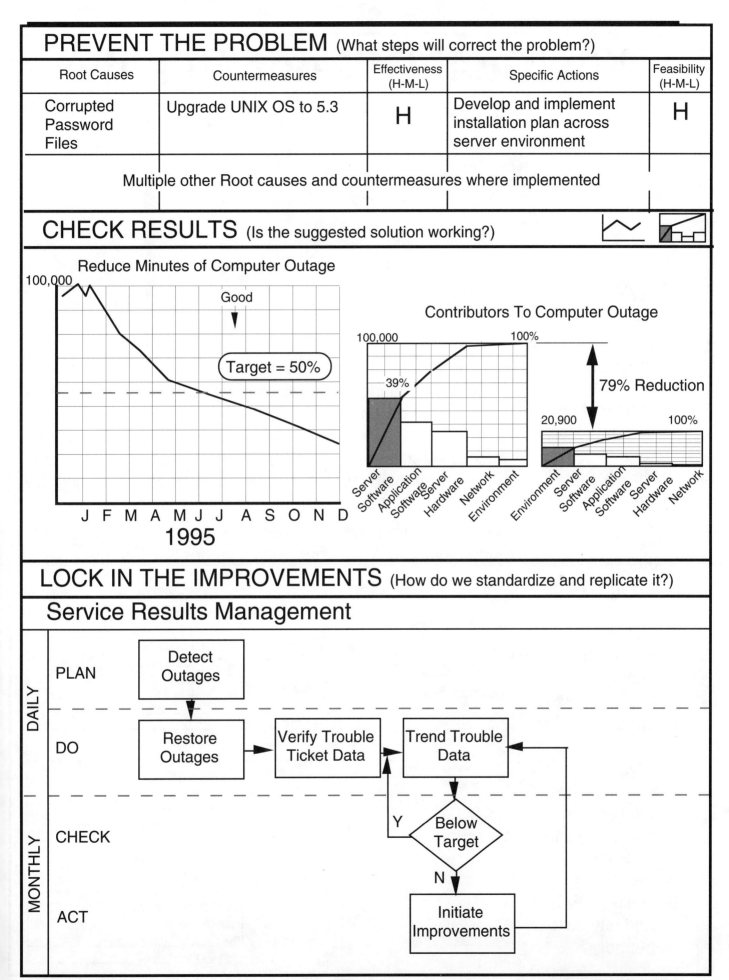

Instructor Guide

The Cost Of Poor Quality
(Waste and Rework)

How?

1. Identify each step in the Fix-it process.
2. Assign a time in minutes to each task and loaded rate.
3. Identify any material costs associated with each step.
4. Identify any external costs of this failure.
5. Identify an lost opportunity, asset, or business costs.
6. Set a target for reducing the error (e.g., 50%)
7. Estimate the total cost of achieving this level of prevention.
8. Evaluate ROI and payback period.

The cost of creating and sustaining high quality processes is far less than the cost of waste and rework incurred when systems fail in operation. Refuse to let your process continue to produce defective results. Learn to "Stop the Line!" Make improvements that will prevent the problem and then continue processing. It's better to prevent a fire than to put one out later.

Evaluate the true costs of a defect or error. This is the foundation of making a business case for the change.

	A	B	C	D	E	F	G
1				Cost of Quality Worksheet			
2	Problem Description: Service Order Errors					Type: Internal	
3	Tasks	Average Hours/ Task	Hourly Rate	Cost of Task	Material Costs	External Failure Cost	Total Cost of Non-Conformance
4	1. Analyze Service Order Error	0.17	$60	$10.00	$3.00	$0.00	$13.00
5	2. Fix Error	0.08	$60	$5.00	$3.00	$0.00	$8.00
6	3. Admin	0.05	$60	$3.00	$0.00	$0.00	$3.00
7	4. Billing Costs Due to Error	0.03	$60	$2.00	$0.00	$0.00	$2.00
8	Total Cost Per Failure						$26.00
9	Service Order Errors/year						221,000
10	1. Lost Opportunity Costs					$0.00	$0.00
11	2. Lost Assets Costs					$0.00	$0.00
12	3. Lost Business Costs					$0.00	$0.00
13	Additional Failure Costs						$0.00
14	Annual Failure Cost						$5,746,000.00
15						Customer or	
16	Basic tasks to fix the problem	Average min/60	Loaded rate	Calculated cost	Expenses	Employee found	Total
17							
18				Return on Investment and Payback			
19				Target Reduction		50%	$2,873,000
20				Prevention Costs			$225,000
21				ROI			$13:$1
22				Payback Period (days)			17

© 2003 Jay Arthur

The Cost Of Poor Quality

Failure Inspection & Prevention

Flops are part of life's menu.
  Rosalind Russell

Any man can make mistakes, but only an idiot persists in his error.
  Cicero

Perfection has one grave defect: it is apt to be dull.
  Somerset Maugham

Many people worry about how much Six Sigma will cost. J. M. Juran suggests the following way of thinking about the costs of quality.

1. If you don't do anything to ensure a quality product, the cost of <u>failures</u> will be too high. They require either <u>rework</u> (e.g., getting a car fixed under warranty) or <u>waste</u> (e.g., food spoilage in a restaurant).

2. Many companies, in a knee-jerk reaction to these failures institute extensive <u>inspection</u> efforts to catch defective products or services before they reach the customer. Again, this requires significant <u>rework</u> and <u>waste</u>. In the worst cases, half the people are involved in inspection and defect removal.

3. Six Sigma companies, on the other hand, focus on <u>preventing</u> defects. If no one puts defects in, then no one has to find them or fix them. This frees everyone to focus on meeting the needs of customers instead of fixing their complaints.

Most companies prefer firefighting to fire prevention. They simply don't do enough process improvement to prevent the waste and rework costs associated with inspection and failure. Your goal should be to find the optimum balance between prevention, inspection, and failure. Remember, less than 4% of the business causes over 50% of the failures. As you move toward six sigma, total costs will decline even more.

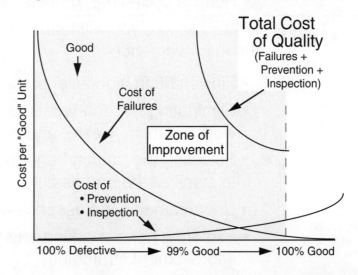

Instructor Guide

Sustain the Improvement

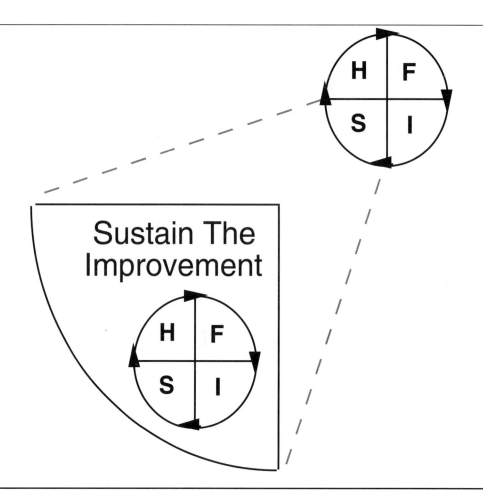

Linkage	In the previous section, you learned how to:
	• Improve the process to deliver breakthrough improvements in speed, quality, and cost
	• Reduce cycle time, defects, waste, and rework
Objectives	In this section, you will learn how to standardize and sustain the improvement by:
	• Flowcharting high level processes and their subprocesses
	• Eliminating the non-valued added time and activities: idle work products, waste, and rework.
	• Selecting the "quality" and "process" indicators necessary to stabilize the process.
	• Understanding the essence of stability and capability.
	• Exploring benchmarking and reengineering as process improvement strategies.

© 2003 Jay Arthur   Six Sigma Instructor Guide

Sustain The Improvement

Process Management

Process Management deals with the ongoing work of an organization. You have internal customers who receive your products or services. By identifying your customers and their needs, you can establish specifications and targets for your work processes and then define measurements to ensure that you can deliver what the customer requires. Continual monitoring will let you know how well you are satisfying your customers and where improvement of your processes is required. Process Management helps you:

Benefits

- achieve consistency in daily work and improved results
- clarify contributions toward achieving customer satisfaction
- systematically improve and control our processes
- maintain the gains achieved through improvement projects
- Identify processes for laser-focused improvement areas
- Provide focus to problem-solving teams
- Sustain the gains from improvement teams
- Assist in training employees
- Multiply the gains from one system to other similar work processes
- Increase employee understanding of what is expected
- Increase communication in the work place

Key Tools

There are three key tools in the process management process:

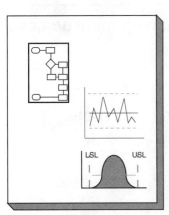

- Flowcharts - to define the process flow

- Line graph (especially control charts) - to measure the stability of the process

- Histogram - to measure the capability of the process

With these three tools you can stabilize and sustain virtually any process.

Instructor Guide

Sustain the Improvement

Process Management

Stabilization

Process Management deals with the ongoing work of an organization. We all have internal customers who receive our products or services. By identifying our customers and their needs, we can establish specifications and targets for our work processes and then define measurements to ensure that we can deliver what the customer requires. Continual monitoring will allow us to know how well we are satisfying our customers and where improvement of our processes is required. Process Management helps us:

- achieve consistency in daily work and improved results
- clarify contributions toward achieving customer satisfaction.
- systematically improve and control our processes
- maintain the gains achieved through improvement projects.

Benefits

- Identifies processes that are directly related to the breakthrough improvement areas
- Provides focus to problem-solving teams especially in steps 1 and 2.
- Helps sustain the gains from improvement teams
    (Step 6 - standardization)
- Assists in training employees
- Multiplies the gains from one system to other similar work processes
- Increases employee understanding of what is expected
- Increases communication in the work place

Key Tools

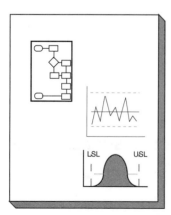

There are three key tools in the process management process:

- Flowcharts - to define the process flow

- Line graph (especially control charts) - to measure the stability of the process

- Histogram - to measure the capability of the process

With these three tools you can stabilize and sustain virtually any processes.

Sustain The Improvement

Process Management

There is always a best way of doing everything.
   -Emerson

After you have found better, faster, and cheaper ways of serving customers, you will want to define and stabilize the new way of doing business. Making existing processes predictable and capable of meeting customer requirements follows the FISH process--Focus, Improve, Sustain, and Honor.

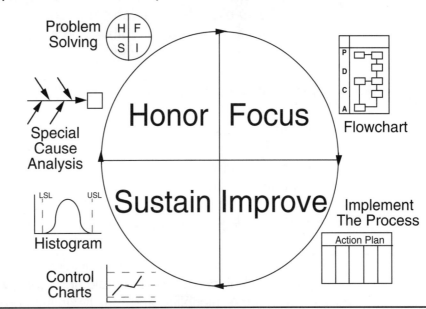

Process

FISH	Step	Activity
Focus	1	Refine the process
	2	Identify the "quality" and "process" indicators
Improve	3	Implement the process and indicators
Sustain	4	Check the process for stability and capability
Honor	5	Recognize, review & refocus

© 2003 Jay Arthur   73   Six Sigma Instructor Guide

Instructor Guide

Sustain the Improvement
Step 1 - Refine The Process

What is a Process?

Simplified, a process is:

Characteristic	Definition	6σ Tool
Repetitive	- hourly, daily, weekly, monthly	
Actions	- step-by-step tasks and activities	
Definable	- observable and documentable	(flowchart)
Inputs	- measurable inputs	(control charts)
Outcomes	- measurable outputs	(control charts)

Flowcharts

Most processes can be diagrammed with only four symbols:

- start/end box
- activity box
- decision diamond
- connecting arrow

Additional symbols can be added as required.

Tips

Creating a flowchart from scratch is like putting together a puzzle: it's best to get all the pieces out on the table and then try to put them in order. To do so requires flexibility and that flexibility comes from using Post-it™ notes (Hint: adhesive on Post-it™ notes is better than on other brands.)

Tar Pits

There are a few tar pits for teams to avoid:

- trying to show too many different kinds of process on one flowchart (e.g., trying to show project management on the same chart as daily operations, or trying to show procurement on the same flowchart as tape operations).

- trying to show too much detail on any one flowchart. Use macro and micro level flowcharts to describe increasing levels of detail.

- using internal "efficiency" indicators rather than external "effectiveness" indicators based on customer requirements.

Sustain The Improvement
Refine The Process

Purpose   Define the <u>existing</u> or improved process as a starting point for stabilization

Flowchart Symbols

A flowchart uses a few simple symbols to show the flow of a process. The symbols are:

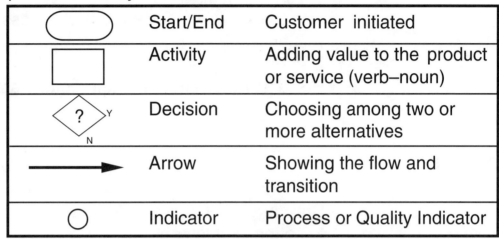

⬭	Start/End	Customer initiated
▭	Activity	Adding value to the product or service (verb–noun)
◇	Decision	Choosing among two or more alternatives
→	Arrow	Showing the flow and transition
○	Indicator	Process or Quality Indicator

The unified process of drawing and shooting was divided into sections: grasping the bow, nocking the arrow, raising the bow, drawing and remaining at the point of highest tension, loosing the shot.
   - Eugen Herrigel

Instead of writing directly on the flowchart, use small Post-it™ notes for both the decisions and activities. This way, the process will remain easy to change until you have it clearly and totally defined. Limit the number of decisions and activities per page. Move detailed subprocesses onto additional pages.

Across the **top** of the flowchart list every person or department that helps deliver the product or service. Along the **left-hand side**, list the major steps in your process: planning, doing, checking, and acting to improve. Even going to the grocery store involves creating a list (plan), getting the groceries (do), checking the list, and acting to get any forgotten item.

Process indicators measure performance <u>during</u> the process. They help find and fix problems before the customer is affected. Put them at critical hand-offs between functions and decision points-especially ones that require error correction.

Quality indicators, measured at <u>after</u> delivery, track customer satisfaction with timeliness, accuracy, and value.

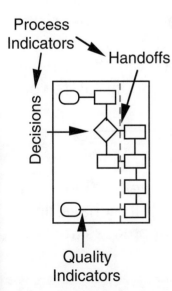

Instructor Guide

Sustain the Improvement
Step 1 - Refine The Process

Process Flowcharts

Process flowcharts extend the flowcharting technique to show "who does what" across the top of the flowchart and the macro steps of the process down the left-hand column. (See next page.) Guidelines for constructing process flow charts include:
- Start with identifying customer needs and end with satisfying them.
- Use the top row to separate the process into areas of responsibility
- Use Post-it™ notes to lay out activities
- Place activities under the appropriate area of responsibility.

Tips

- Use square Post-it™ notes for activities and decision diamonds.
- Draw arrows on any size Post-it™ note to show the flow, top to bottom, left to right.

Flowcharting
- Use smaller Post-it™ notes for process and quality indicators.
- Participants will often offer activities at different levels of detail. As the higher level process flow gets more complex, keep moving sub-processes onto micro process diagrams.

Quality indicators
- Quality indicators which measure how well the process met the customer's requirements go at the <u>end</u> of the process.

Process indicators
- Process indicators which predict how well the process will meet the requirements are most often placed at: 1) hand-offs between functional groups and 2) at decision points to measure the amount work flowing in each direction (this is most often useful for measuring the amount of rework required).

EXERCISE

Purpose: To develop a flowchart
Agenda:
1. Closely coach participants to develop their own individual or team macro flowchart. If they have time, have them do the most important subprocess from the macro flow.

Limit: 45 minutes

Sustain The Improvement
Refine The Process

Step \ Who	Customer				
		FLOWCHART			
Plan	Request for product or service				
Do					
Check					
Act	Delivery of product or service				

Instructor Guide

Sustain the Improvement
Monitor the Indicators

Indicators

Indicators
- **Stable**
- **Capable**

Each customer requirement must have a corresponding measurement or indicator, typically a line graph showing trends over time. These indicators are vital to breakthrough improvement. For a business to be successful, processes must be stable and capable of meeting customer requirements.

Process Performance

- Stable--a stable process will predictably deliver what the customer wants.

  Example: Do you have restaurants you patronize because you consistently get what you want in the time allotted?

- Capable--a capable process meets the customer's expectations 100% of the time.

  Again: how long do you patronize restaurants where the food is inadequate, checks are wrong, service is poor?

Track the Quality Indicators

Quality Indicators

To each thing belongs its measure.
- Pindar

Examples

Timeliness is best in all matters
- Hesiod

Most measures of customer requirements can be easily expressed as a line graph (showing periodic variation over time) or a histogram showing a snapshot of overall performance:

Requirement:
I want an accurate paycheck

Measurement:
Paycheck errors

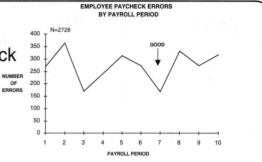

Requirement:
Don't waste my time
Do it in an hour or less

Measurement:
Time to install service

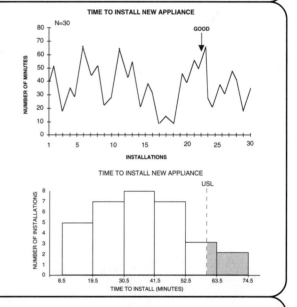

Requirement:
Be efficient
Don't waste food

Measurement:
Cost of food spoilage

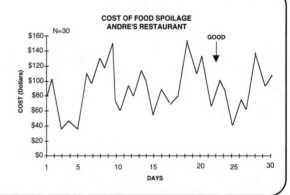

Instructor Guide

Sustain the Improvement
Monitor the Indicators

Exercise (Optional)

Purpose: Graph the improvement indicators

Agenda:

- In small groups, have participants select one indicator for good, fast, and cheap. Using real or best guess data, have participants plot the current performance.
- Which direction is good, up or down, increase or decrease? (e.g., reduce defects, time, cost).
- Add a line showing the target for improvement

Limit: 20 minutes

Purpose: Indicators

Agenda:
- Line graph
- Good arrow

Limit: 20 minutes

Track the Quality Indicators

Quality Indicators

TIP: Indicators (of the customer's requirements) are measured <u>after</u> the product or service is delivered.

Quality Indicators

Who collected the data?

When was it collected?

Where?

What formula used?

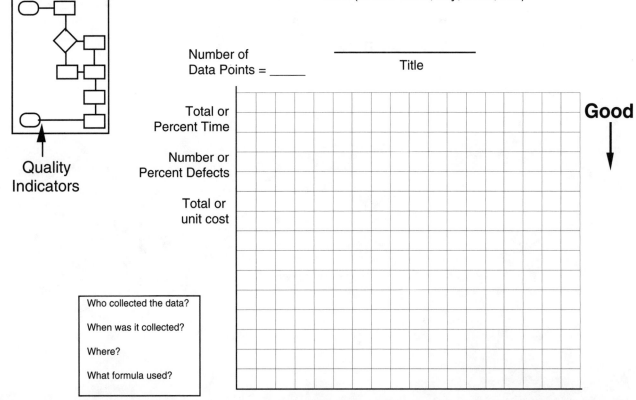

Instructor Guide

Sustain the Improvement
Step 2 - Identify the Indicators

Exercise

Purpose:
   Indicators

Agenda:
   • Quality
   • Process

Limit: 30 minutes

For improvement efforts to be successful, they must focus on the customer's requirements and ways to measure them—defects, time, or cost. Earlier, in planning, we developed indicators based on customer requirements. These are usually the "quality" indicators measured after delivery of the product or service. Now we need to identify the hand-off and decision points where "process" indicators can be measured to predict the performance of the process.

Purpose: Develop the process indicators

Agenda:

   • In small groups, have participants identify one "quality" indicator based on customer requirements.

   • Using the process flowchart, have participants identify one or two places in the process where a measurement indicator would reliably predict the results (quality indicator). The number errors corrected during the process, for example, will predict the quality of the final product (lots of errors probably means a poor product).

Limit: 30 minutes

Identify the Process Indicators

Process Indicators

The measurements of customer requirements usually occur after the end product or service is delivered. To ensure that customers get what they want, we have to set up a system of early warning indicators that will predict whether or not the process will deliver what the customers want. Like the quality indicators, these predictive indicators will need to measure defects, time, and cost <u>inside the process</u>.

Where to Measure

Often, the easiest points to measure:

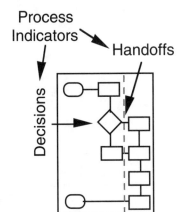

- **process indicators** are at the key handoffs (to measure time or missed commitments) or decision points (to measure defects and rework). If, for example, you were trying to <u>predict</u> how long it would take to get to work, the number of red lights or average highway speed could <u>predict</u> the total commute time.

- **quality indicators** (e.g., total commute time) <u>after</u> the product or service has been delivered. Looking back at the main flowchart, at what points could you most easily take measurements that would predict whether the process will be able to deliver what the customers want?

Examples

"Quality" Requirements Indicator	"Process" Early Warning Indicator
Percent defective	Amount of rework per step Number of defects per step
Missed Commitments	Time per process step Delay (idle and rework time)
Value	Cost of waste and rework
Paycheck errors	Timesheet errors % timesheets late
Appliance installation time	Old appliance removal time
Cost of food spoilage	Number of customers Perishable food ordered

Instructor Guide

Sustain the Improvement
Step 3 - Track the Indicators

Exercise (Optional)

Purpose: Graph the process indicators

Agenda:

- In small groups, have participants select one indicator for good, fast, and cheap. Using real or best guess data, have participants plot the current performance.
- Which direction is good? (Reduce defects, time, cost).

Limit: 20 minutes

Purpose: Indicators

Agenda:
- Line graph
- Good arrow

Limit: 20 minutes

Track the Process Indicators

Process Indicators

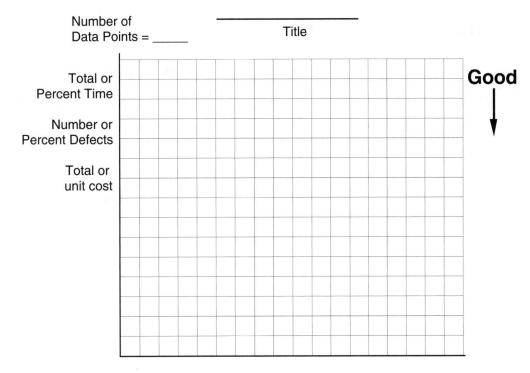

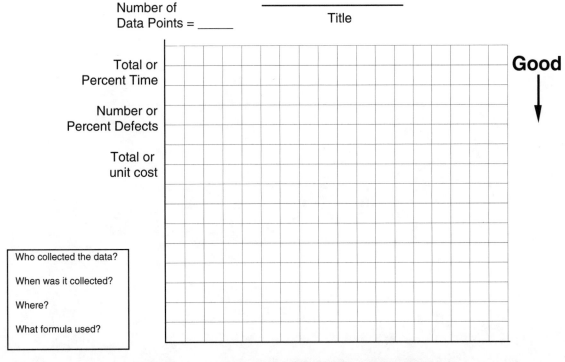

Instructor Guide

Sustain the Improvement
Sampling

Sampling

Sampling is a science for reducing the cost of <u>inspecting</u> materials, products and services. Let participants know it's possible and offer to get or give them expert help when they need it.

Check Stability
Sampling

Purpose — Minimize the cost of collecting the data.

Sampling

By a small sample may we judge the whole piece.
- Cervantes

To begin managing a process, you will need a minimum of 20 data points for each indicator. To collect data cost effectively, you will have to understand the basic concepts of sampling. If you are only producing ten widgets a day, then it is fairly easy to look at all of them (the *total population* of created widgets). If you produce 100,000 widgets a day, however, you will want to look at a small sample and *draw conclusions* about the entire population from the sample:

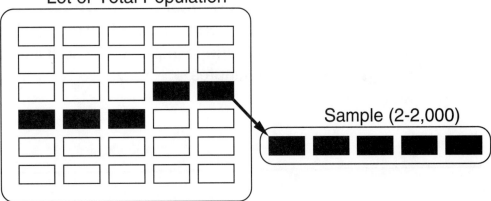

In the population, if there is a . . .	then . . .
large number that are measured	
manually	take a sample
mechanically	use the total population
small number	use the total population

Sample Size — When samples are taken, they should be the same size: If you check five widgets in this sample, then you will want to check five widgets *every* time you sample. If you use the total population, it can vary from period to period or be a constant size. In general, the sample size will vary based on: the acceptable level of quality desired, the size of the "lot" you're inspecting, the type of sampling done--single, double, multiple, and the level of inspection. Higher quality requires larger samples.

Instructor Guide

Sustain the Improvement
Step 4 - Check Stability

Stability

> **Key Point**
> Stable
> =
> Predictable

A stable process produces <u>predictable results consistently</u>.

Example: Ask participants: "How long does it take you to commute to work each morning?" Is it a "stable" process? What kinds of events could make it unstable? Is it capable of meeting your requirements for commute time, family time, etc.?

Exercise

Purpose: To evaluate stability
Agenda:
1. Review one quality indicator reflecting a customer requirement.
2. Based on real or intuitive data, is the process stable? Does it produce consistent results? If not, what would need to be done to improve the consistency of results?

Limit: 15 minutes

Stabilize the Process
Understanding Stability

Stability

A stable process produces <u>predictable results consistently</u>. Stability can be easily determined from control charts. The upper control limit (UCL) and lower control limit (LCL) are <u>calculated</u> from the data.

Example

How long does it take you to commute to work each morning?

| Stable = Predictable |

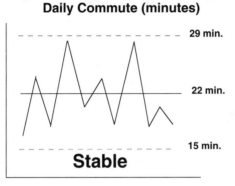

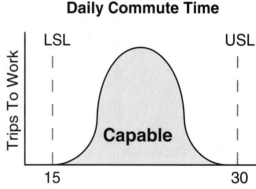

Your Specification Limits:
1. Get to work in 30 minutes or less.
2. Get to work safely (no faster than 15 minutes).

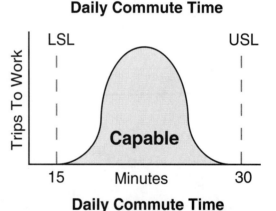

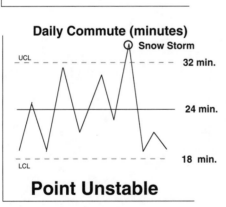

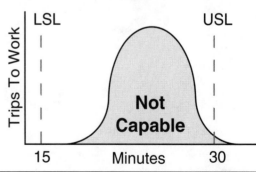

Stability and Capability

A process does not have to be stable to be capable of meeting the customer's requirements. Similarly, a stable process is not necessarily capable. A managed process must be both stable <u>and</u> capable. Interpreting stability with control charts and capability with histograms will be discussed in more detail on the following pages.

© 2003 Jay Arthur — Six Sigma Instructor Guide

Instructor Guide

Sustain the Improvement
Step 4 - Interpreting the Indicators

Indicators

> **Indicators**
> - **Stable**
> - **Capable**

Each customer requirement must have a corresponding measurement or indicator, typically a line graph showing trends over time. These indicators are vital to breakthrough improvement. For a business to be successful, processes must be stable and capable of meeting customer requirements.

Process Performance

- <u>Stable</u>--a stable process will predictably deliver what the customer wants.

    Example: Do you have restaurants you patronize because you consistently get what you want in the time allotted?

- <u>Capable</u>--a capable process meets the customer's expectations 100% of the time.

    Again: how long do you patronize restaurants where the food is inadequate, checks are wrong, service is poor?

Tip: Participants need to know more about how to interpret charts than how to construct them. There are lots of tools to do the calculations. Teach them how to recognize patterns of instability. Relate each example back to daily commute times.

Check Stability
Interpreting The Indicators

Purpose — Verify that the process system is stable and can predictably meet customer requirements

Variation

You cannot step twice into the same river.
   Heraclitus

A stable process produces <u>predictable results</u>. Understanding variation helps us learn how to predict the performance of any process. To ensure that the process is stable (i.e., predictable) we need to develop "run" or "control" charts of our indicators.

How can you tell if a process is stable? Processes are never perfect. *Common* and *special causes* of variation make the process perform differently in different situations. Getting from your home to school or work takes varying amounts of time because of traffic or transportation delays. These are <u>common causes</u> of variation; they exist every day. A blizzard, a traffic accident, a chemical spill, or other freak occurrence that causes major delays would be a <u>special cause</u> of variation.

In the 1920s, Dr. Shewhart, at Bell Labs, developed ways to evaluate whether the data on a line graph is common cause or special cause variation. Using 20-30 data points, you can determine how stable and predictable the process is. Using simple equations, you can calculate the average (center line), and the upper and lower "control limits" from the data. 99% of all *expected* (i.e., common cause variation) should lie between these two limits. Control limits are not to be confused with specification limits. Specification limits are defined by the customer. Control limits show what the process can deliver.

Example

Your Requirements:
1. Get to work fast!
2. Get to work safely.

Daily Commute (minutes)

— 29 min.
— 22 min.
— 15 min.

Stable

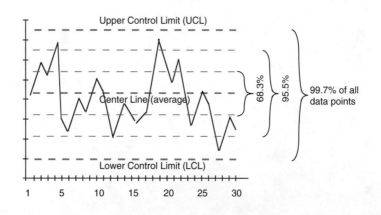

© 2003 Jay Arthur   82   Six Sigma Instructor Guide

Instructor Guide

Sustain the Improvement
Step 4 - Interpreting the Indicators

Exercise

Purpose: Interpret a control chart

Agenda:

- In small groups, give participants two control charts. Have them analyze the charts for stability.

Limit: 20 minutes

Purpose: Control Chart

Agenda:
- Stability

Limit: 20 minutes

Check Stability
Interpreting The Indicators

Special Cause Variation

Processes that are "out of control" need to be stabilized before they can be improved using the problem-solving process. Special causes, require immediate cause-effect analysis to eliminate the special cause of variation.

Evaluating Stability

The following diagram will help you evaluate stability in any control chart. Unstable conditions can be any of the following:

Point Unstable

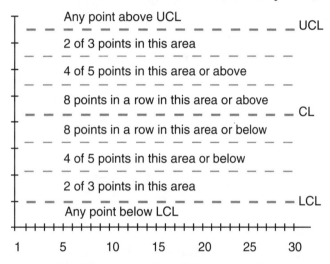

Points and Runs

Any point outside the upper or lower control limits is a clear example of a special cause. The other forms of special cause variation are called "runs." Trends, cycling up and down, or "hugging" the center line or limits are special forms of a run.

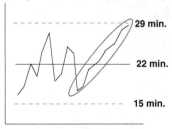

Unstable Trend

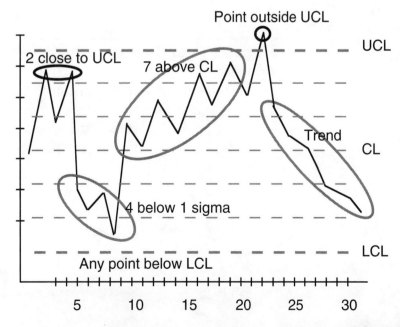

© 2003 Jay Arthur

Instructor Guide

Sustain the Improvement
Step 4 - Check Capability

Capability

A **capable** process <u>meets the customer's requirements 100% of the time</u>.

> **Key Point**
> Capable
> =
> Meets Customer Requirements 100% of the Time

Tip: Participants always try to confuse control limits (control charts) and specification limits (histograms). Control limits are <u>calculated</u> from the data. The upper (USL) and lower (LSL) <u>specification</u> limits are determined from the customer's requirements.

Tip: Determining capability from data requires that you know what kind of data you're using. Participants try to get confused about the two types of data: <u>counted</u> data (defects, missed commitments, etc.) and <u>measured</u> data (time, length, weight, money). Counted data is binary (i.e., yes/no, on/off); either you missed a commitment or you didn't. Measured data is infinitely divisible (e.g., micrometer for width or decimals for money . . . especially in your bank account).

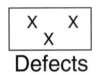
Defects

The <u>capability</u> of **counted** data (e.g., defects--indivisible integers only) is <u>zero defects</u>. Customers hate defects--outages (USL = LSL = 0).

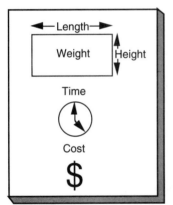

The <u>capability</u> of **measured** data (e.g., time, money, age, length, width, weight, etc.) is determined using the customer's specifications and a histogram (see below).

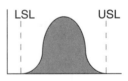

Exercise

Purpose: To evaluate capability
Agenda:
1. Review one quality indicator reflecting a customer requirement.
2. Based on real or intuitive data, is the process capable? Does it meet customer requirements 100% of the time? If not, what would need to be done to improve the process to meet requirements?

Limit: 15 minutes

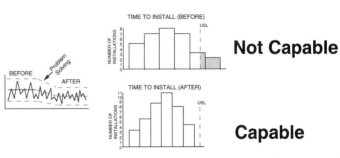

Stabilize the Process
Understanding Capability

Capability

A **capable** process meets the customer's requirements 100% of the time. The upper (USL) and lower (LSL) specification limits are determined from the customer's requirements.

Defects

The capability of **counted** (i.e., attribute) data like defects--indivisible integers only-- is zero defects. Customers hate defects--outages (USL = LSL = 0).

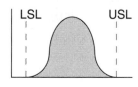

The capability of **measured** (i.e., variable) data like time, money, age, length, width, and weight is determined using the customer's specifications and a histogram (see below).

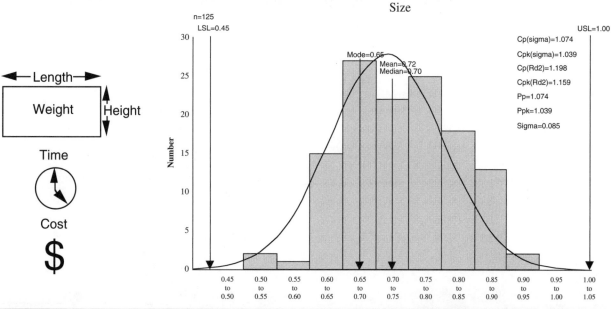

Problem Solving

Is the process capable? If not, what improvement activities are required to make the process both stable and capable?

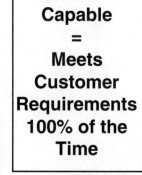

Capable = Meets Customer Requirements 100% of the Time

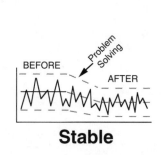

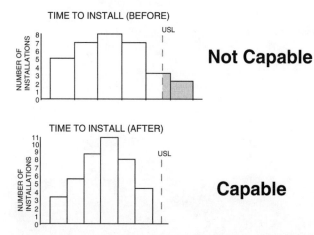

Sustain the Improvement
Understanding Capability

Goal Post vs Target

As a football society, it used to be "good enough" to put the ball through the uprights, but not any more. Rather than focusing on how to make products "fit" between the upper and lower specification limits, world-class companies focus on how to minimize variation around a target value. This helps eliminate waste and rework both during production and in use.

Ask participants: have you ever tried to put together a child's toy during the holidays only to find that something won't fit and you have to take it back? When the upper "goal post" for the peg exceeds the lower goal post for the hole, neither will fit and you have a problem.

What are some targets that might be appropriate for your organization?

Stabilize the Process
Understanding Capability

Goal Post vs Target

Traditional thinking took a "goal post" attitude toward process capability. When the customer defines an upper and a lower specification limit for a product or service-whether it's the diameter of a shaft or the time in line at a fast food restaurant, all points within the two limits are considered "good." But Taguchi suggested that there is a loss incurred by society as these products and services move away from their target value. This cost increases exponentially with distance from the target value:

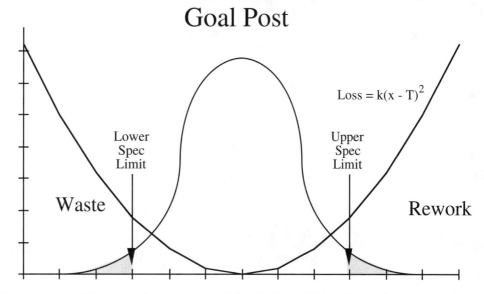

Target thinking encourages minimization of variation, with a resulting savings based on Taguchi's "loss function."

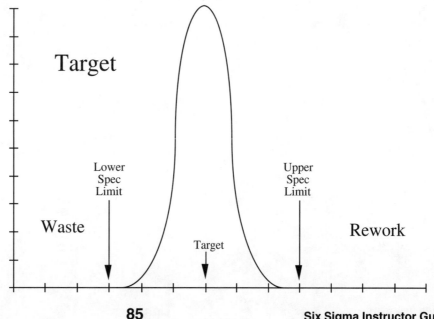

Instructor Guide

Sustain the Improvement
Understanding Capability

Several **capability indices** help evaluate process variation, Cp, Cpk, Pp, and Ppk. The Cp indexes measure inherent variation in stable processes, and the Pp indexes measure total variation.

Inherent process variation, due to common causes *only*, is estimated by the average of the range within a sample times d_2: Rd_2.

Total process variation, which includes special causes, is estimated by the sample standard deviation: σ_s.

Cp
    Cp is the capability index that evaluates the spread of the process. Cp answers the question: If the process were centered, would it fit between the goal posts.

Cpk
    Cpk is the capability index that evaluates how centered the process is between the goal posts.

Pp
    Pp is the process performance index that evaluates the *total* spread of the process.

Ppk
    Ppk is the process performance index that evaluates how centered the process is between the goal posts.

Note: Some texts use the formula for Pp as the formula for Cp, and Ppk for Cpk. The difference is inherent variation in stable processes, vs total variation.

Stabilize the Process
Understanding Capability

Is the process capable? Cp, the process capability index, is the 6σ range of the variation within the process. Cp doesn't care if the process is centered over the target or is off-center, all it cares about is whether or not the data points would fit within the upper and lower specification limits.

Cp

Cp=(USL-LSL)/6σ – Capability index with out centering

where σ is estimated by $\bar{R}/d_2$ when the process is *statistically stable*.
$\bar{R}$ is the average of the ranges in samples
d_2 is a constant based on the sample size.

Cpk

Cpk, on the other hand, accounts for process centering.

Cpk=Minimum of $(USL-\bar{\bar{X}})/3\sigma$ or $(\bar{\bar{X}}-LSL)/3\sigma)$

Capable Cp≥1, Cpk≥1

Typically, when Cp and Cpk are over 1.0, the process is capable. In the example below, $Cp_{Left}=Cp_{Right}$, while $Cpk_{Left} > Cpk_{Right}$, because of process centering.

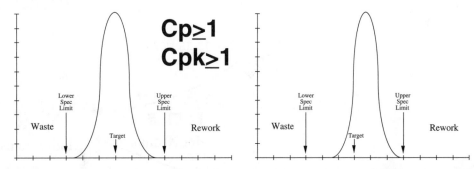

A Cpk value less than zero indicates that most of the points fall outside the specification limits. Note that Cp, in this example is still greater than 1.

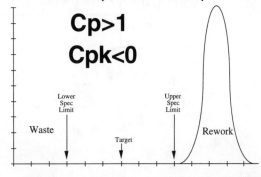

Instructor Guide

Sustain the Improvement
Step 4 - Choosing a Chart

Indicators

The type of control chart to use depends on the size of the sample and the type of data (see below). That's all they need to know. Let them know that you can help them when they need to select and construct a chart. Be their expert. Most people are scared to death of math and graphs.

Type of Chart
- Sample Size
- Type of Data

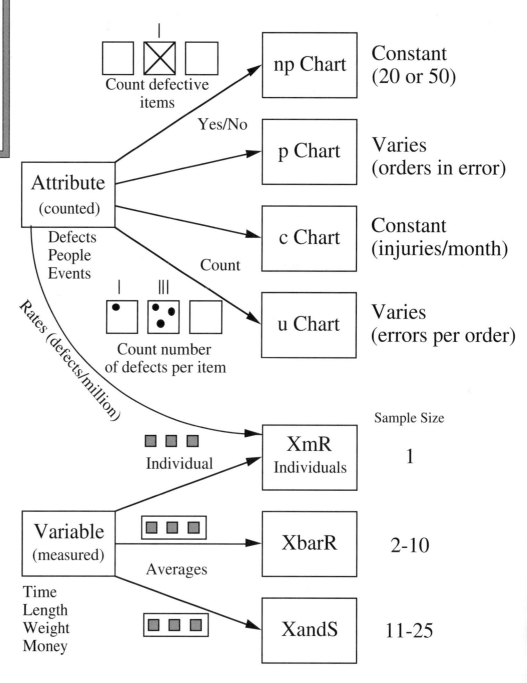

Check Stability
Choosing a Chart

Types of Charts

Next, you have to know what to collect about each widget. Do you need to know: how long it takes to deliver a product or service, the number of defects per product, or the cost of waste or rework? Time, cost, length, and weight are known as *variable* data. Counting the number of defects or defective items gives *attribute* data. The type of data (attribute or variable) and the size of the sample taken (1, 2-10, or total) will determine the type of graph used to measure the process.

Choosing a Chart

Attribute data

Variable data

		Sample Size		
Type of data		1	Same Size	Varies
X = 1 Defective (pp. 36-37)			np (2-total)	p (total)
X X X = 3 Defects (pp. 38-39)			c (2-total)	u (total)
Length, Weight, Height, Time, Cost (pp. 40-41)		XmR	$\bar{X}R$ $\bar{X}$andS (size 2-25)	
Y Axis		Number	Number	Percent

Drawing the Chart

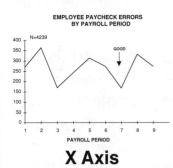

X Axis

Regardless of sample size, each of these charts can be drawn as a *line graph.* **Tip:** If the math seems scary, start with line graphs or get the 6σ Macros for Microsoft® Excel.
- The **X axis** (horizontal) shows how often the data is collected (daily, hourly, weekly, periodically).
- The **Y axis** (vertical) shows:
    - the number or percent defective (c, np, p, u)
    - the time, cost, length, weight, etc. (XR charts)

Then, based on the type of data and sample size, you can calculate the upper and lower control limits (UCL, LCL) and center line (CL) that will make it possible to evaluate process stability. The next few pages show how to calculate and interpret the limits.

© 2003 Jay Arthur   Six Sigma Instructor Guide

Step 4 - Check Stability
np and p charts

p and np Charts
(Attribute data)

Defective

The p and np charts will help you evaluate process stability when counting the number or fraction <u>defective</u>. Examples might include: the number of defective circuit boards, meals in a restaurant, teller interactions in a bank, invoices, or bills.

The np chart is useful when it's easy to count the number of defective items and the sample size is always the same. The p chart is used when the sample size varies: the total number of circuit boards, meals, or bills delivered varies from one sampling period to the next. In the p chart below, the number of defective paychecks varies with the number of employees in each pay period.

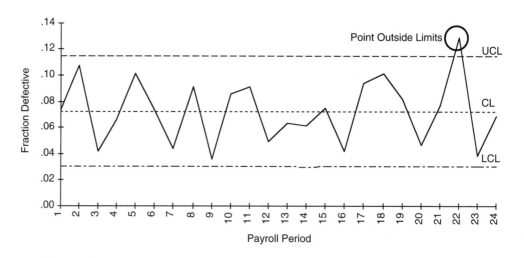

Stability

Given this information, we would want to investigate why the 22nd payroll period was "out of control." Otherwise, this chart, and therefore this process, look stable.

Capability

A fully capable process delivers <u>zero defects</u>. Although this may be difficult to achieve, it should still be our goal. Once we resolve the out-of-control point, we could use the problem solving process to begin to eliminate the common causes of defective paychecks. What are the most common types of paycheck errors? Why do they occur? What are the root causes of these paycheck errors?

© 2003 Jay Arthur   Six Sigma Instructor Guide

Step 4 - Check Stability
np and p charts

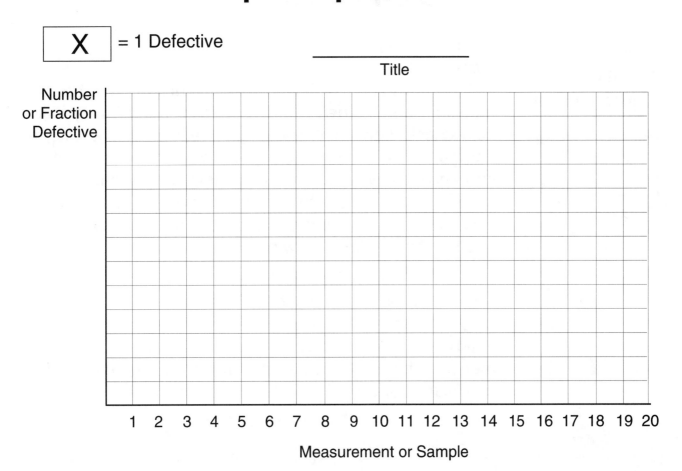

p Chart
UCL: $\bar{p} + 3*\text{sqrt}(\bar{p}*(1-\bar{p})/n_i)$
CL: $\bar{p} = \Sigma p_i/\Sigma n_i$
LCL: $\bar{p} - 3*\text{sqrt}(\bar{p}*(1-\bar{p})/n_i)$

np Chart
$n\bar{p} + 3*\text{sqrt}(n\bar{p}*(1-n\bar{p}/n))$
$n\bar{p} = \Sigma np_i/k$
$n\bar{p} - 3*\text{sqrt}(n\bar{p}*(1-n\bar{p}/n))$

Step 4 - Check Stability
c and u charts

c and u Charts
(Attribute data)

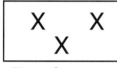
Defects

The c and u charts will help you evaluate process stability when there can be more than one defect per unit. Examples might include: the number of defective elements on a circuit board, the number of defects in a dining experience--order wrong, food too cold, check wrong, or the number of defects in bank statement, invoice, or bill. This chart is especially useful when you want to know how many defects there are not just how many defective items there are. It's one thing to know how many defective circuit boards, meals, statements, invoices, or bills there are; it is another thing to know how many defects were found in these defective items.

The c chart is useful when it's easy to count the number of defects and the sample size is always the same. The u chart is used when the sample size varies: the number of circuit boards, meals, or bills delivered each day varies. The c chart below shows the number of defects per day in a uniform sample.

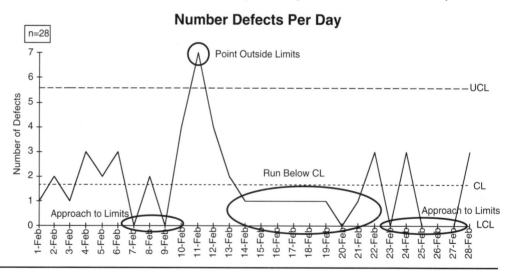

Stability

Given this information, we would want to investigate why February 11th was "out of control." We would also want to understand why we were able to keep the defects so far below average in the other circled areas. What did we do here that was so successful?

Capability

A fully capable process delivers <u>zero defects</u>.

Step 4 - Check Stability
c and u charts

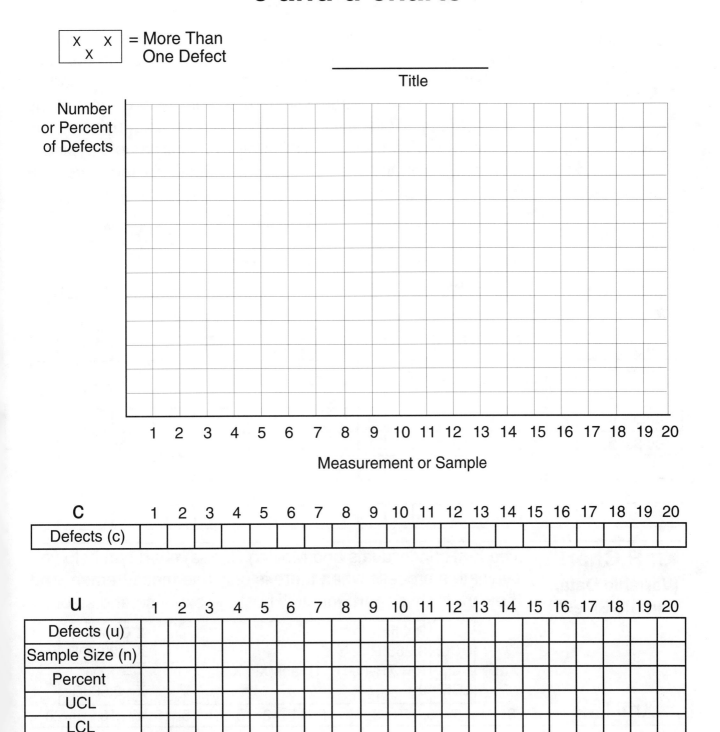

C Chart
UCL: $\bar{c} + 3*sqrt(\bar{c})$
CL: $\bar{c} = \sum c_i/n$
LCL: $\bar{c} - 3*sqrt(\bar{c})$

U Chart
$\bar{u} + 3*sqrt(\bar{u}/n_i)$
$\bar{u} = \sum u_i / \sum n_i$
$\bar{u} - 3*sqrt(\bar{u}/n_i)$

Step 4 - Check Stability
XandR Charts

$\overline{X}R$ Chart
(Variable data Sample Size=5)

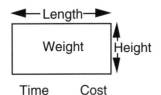

The $\overline{X}R$ chart can help you evaluate the cycle time for almost any process: making a widget, answering a customer call, seating a customer, delivering a pizza, or servicing an appliance. This chart is especially useful when you do this many times a day. Collecting the data could be expensive if you measured every time you did it. Using a small sample (typically five and as many as 25) you can effectively measure and evaluate the process.

		R Chart	(For Sample Size=5)	X Chart
$\overline{X}R$	UCL:	$2.114*\overline{R}$		$\overline{\overline{X}} + .577*\overline{R}$
	CL:	$\overline{R} = \Sigma R_i/k$	$R = Max(X_i) - Min(X_i)$	$\overline{\overline{X}} = \Sigma \overline{X}_i/k$
	LCL:	0		$\overline{\overline{X}} - .577*\overline{R}$

$\overline{X}R$

	1	2	3	4	5	6	7	8	9	10	11	12	13	14	15	16	17	18	19	20
Sample 1																				
Sample 2																				
Sample 3																				
Sample 4																				
Sample 5																				
Total																				
Average ($\overline{X}$)																				
Range (R)																				

XmR Chart
(Variable Data, Sample Size=1)

The XmR (Individuals and Moving Range) chart can help you evaluate a process when there is only one measurement and they are farther apart: monthly postage expense and so on.

		R Chart	X Chart
XmR	UCL:	$3.268*\overline{R}$	$\overline{X} + 2.660\overline{R}$
	CL:	$\overline{R} = \Sigma R_i/(k-1)$ $R=abs(X_i - X_{i-1})$	$\overline{X} = \Sigma X_i/k$
	LCL:	0	$\overline{X} - 2.660\overline{R}$

XmR

	1	2	3	4	5	6	7	8	9	10	11	12	13	14	15	16	17	18	19	20
X Value																				
Range (R)																				

Calculate, plot, and evaluate the <u>range chart first</u>. If it is "out of control," so is the process. If the range chart looks okay, then calculate, plot, and evaluate the X chart.

Step 4 - Check Stability
XR Charts

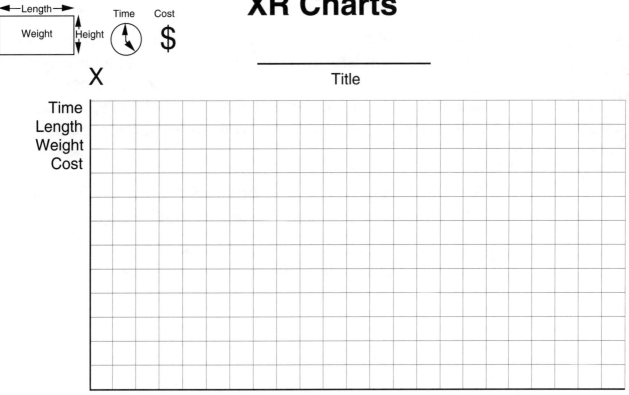

Measurement Interval (minute, hour, day, week, etc.)

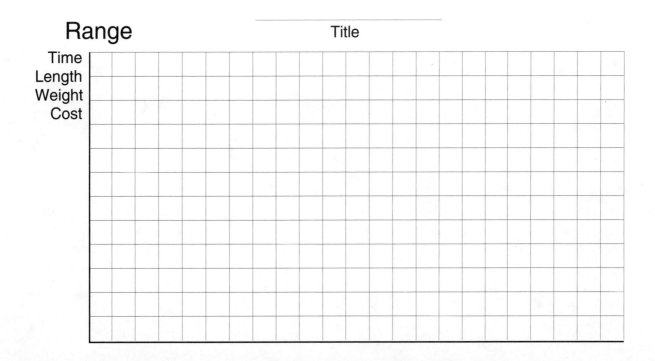

Measurement Interval (minute, hour, day, week, etc.)

Instructor Guide

Sustain the Improvement

Special Cause Analysis

Teams sometimes forget that you can use tools in more than one place. The Ishikawa diagram is a great tool for analyzing special causes of variation in a control chart. These include points outside the upper and lower control limits, trends, and other special causes identified when *interpreting the indicators*.

Ask participants: What might be some special causes of longer than expected commuting times? Storms? Accidents?

Then ask: What are some special causes of delays or defects in the business problem at hand?

Act To Improve
Special Cause Analysis

Purpose

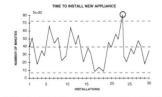

Improve process stability

If the process is <u>not stable</u>, we can use the quality improvement problem solving tools, especially the Ishikawa Diagram, to identify the root causes of the instability, remove them, and make the process stable, repeatable, and predictable.

Special Cause Analysis

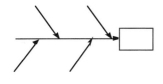

For every thousand hacking at the leaves of evil, there is one striking at the root.
  -Thoreau

Every why has a wherefore.
  - Shakespeare

A good garden may have some weeds.
  Thomas Fuller, M.D.

1. To identify root causes, use the fishbone or Ishikawa diagram. Put a problem statement about the special cause of variation in the head of the fish and the major causes at the end of the major bones. Major causes include:
   - Processes, machines, materials, measurement, people, environment
   - Steps of a process (step1, step2, etc.)
   - Whatever makes sense

2. Begin with the most likely main cause.

3. For each cause, ask "Why?" up to five times.

4. Circle one-to-five <u>root</u> causes (end of "why" chain)

5. Verify the root causes with data (Pareto, Scatter)

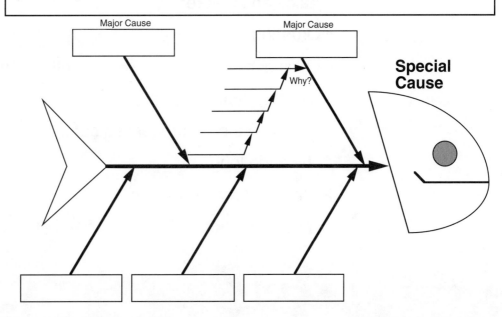

© 2003 Jay Arthur   94   Six Sigma Instructor Guide

Instructor Guide

Sustain the Improvement
Act To Improve

Problem Solving

If a process is producing products that are outside of the specification limits or producing too many defects per million, then it's time to return to the problem solving process. This is pretty straightforward in manufacturing, but seems a little different in service industries.

In healthcare, for example, hospitals are measuring things like:

- Patient falls
- Secondary infections
- Complications
- Prescription errors
- Time from diagnosis to medication
- Length of stay

As you might imagine, every patient can benefit from reduction or elimination of these issues. The first four are attribute charts. Patient's would like those to be zero. The second two are variable charts; patients would like those to be a minimal as possible.

No matter what kind of business you are in, there is always:

- Sales and marketing
- Ordering
- Fulfillment of the order (whether it's a car or healing)
- Billing and payment processing

Any of these core processes can:

1. take too long
2. have too many defects
3. cost too much

Act To Improve
Problem Solving

Purpose — Improve process capability

Common Cause Analysis — If the process is <u>not capable</u> of meeting the customer's needs and expectations, then we can use the problem solving process to begin to analyze and remove the so called "common causes" of variation in the process.

The problem-solving process follows the FISH cycle step-by-step to ensure continuous, never-ending improvement:

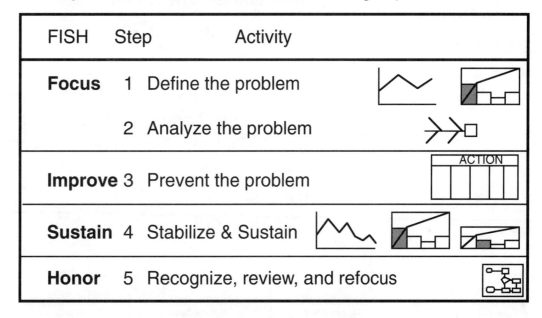

FISH	Step	Activity
Focus	1	Define the problem
	2	Analyze the problem
Improve	3	Prevent the problem
Sustain	4	Stabilize & Sustain
Honor	5	Recognize, review, and refocus

For more detail, see *Double Your Quality*

Stability and Capability *after* Problem Solving

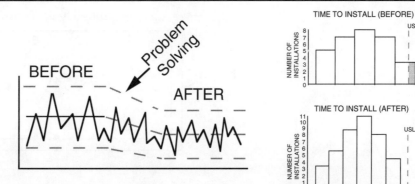

Instructor Guide

Sustain the Improvement
Deploy Monitoring System

Exercise (Optional)

Keeping track of all of the improvements would be complex if you didn't use some form of "flag" system. Since each customer requirement has a corresponding measurement or indicator--a line graph, each goal and its contributing targets can be easily displayed using a tree diagram.

Purpose:
Flag system

Agenda:
- Tree Diagram
- Line Graphs

Limit: 20 minutes

Flags
- Green--on course to meet objective.
- Yellow--target in jeopardy, needs management attention.
- Red--target missed, needs immediate attention.

Purpose: Create a flag system

Agenda:

- In small groups, have participants select one target for improvement that has multiple contributors. Using real or best guess data, have participants create a flag system showing current performance.

Limit: 20 minutes

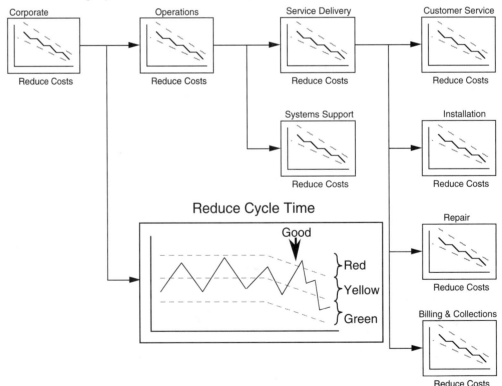

Flag System for Indicator Rollup

© 2003 Jay Arthur   I-96   Six Sigma Instructor Guide

Track Performance
Deploy Monitoring System

Purpose To visually link targets and measurements throughout the business, and to show progress.

Playing catchball creates the targets and means. The flag system links targets and measurements across an organization. The flag system shows how each group will contribute to the achievement of the overall goal. The flag system is a tree diagram that contains graphs of each group's performance.

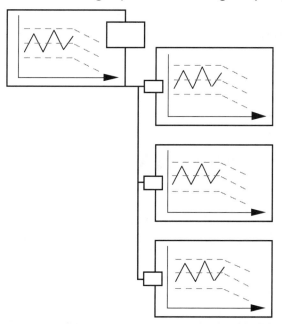

Flag System

FISH	Step	Activity
Focus	1	Identify the gap between current performance and that required for customer satisfaction.
	2	Identify the targets for each group.
Improve	3	Plot the overall improvement graph.
	4	Plot the contributing graphs for each target.
	5	Display and update these graphs monthly.
Sustain	6	Review movement toward targets.
Honor	7	Refocus improvement efforts as required.

Instructor Guide

Benchmarking
Improve The Process

Adopt or Adapt a Better Process

Benchmarking is simply a way to borrow from the best. One data center, for example, may already have a best practice that other centers could use to be more effective. Borrow it!

Key Tools

There are two key tools in the benchmarking process:

- Process Flowcharts - to define the current and benchmarked process

- Line Graphs - to measure and compare the performance of both processes.

With these two tools you can evaluate and adapt any superior process to be your own.

Benchmarking
Improve and Sustain

Purpose — Adopt or adapt best practices

Most people believe that given the resources they are allotted and pressure that is on them, they are doing a good (if not excellent) job. Yet they have no real evidence to justify this belief; nor can they state what constitutes the best performance. Benchmarking adds objectivity to this situation and is a systematic means to become the best. Benchmarking is applicable to innovation, reengineering, and all aspects of process improvement.

Benchmarking Process

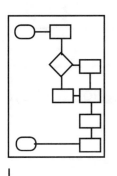

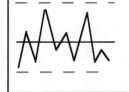

1. Identify the firm's key success factors and benchmark the processes most essential to its success. Assemble a team to conduct the benchmarking and implement the findings.

2. Use process management to determine the firm's baseline: what steps are required, the time and cost of each step, the staffing needed, the materials and machines used, the resulting quality, and the waste or rework.

3. Identify the sources of information (people, articles, etc.) on the best-in-class processes--either internal or external.

4. Visit the groups or firms that seem most promising from the screening. Compare their performance with your own. Set performance targets to close any gaps. Present the findings to senior management and key stakeholders.

5. Create an action plan for achieving the goals.

6. Implement the improved process and monitor progress.

7. Recalibrate the benchmarks each year.

Instructor Guide

Process Reengineering
Improve and Sustain

Reengineering

Where the laser-focused improvement process deals in improvements, reengineering focuses on transformations employing technology. FISH--Focus, Improve, Sustain, and Honor--can be a *revolutionary* as well as an *evolutionary* method of improvement. A balance of innovation and incremental improvement is essential to long-term success.

If processes have undergone little change in the last five to ten years, then they are likely candidates for reengineering. Technological innovations are expanding radically. Yesterday's impossibility is today's simple, cost-effective solution.

Process

The reengineering process follows the FISH cycle to ensure continuous, never-ending improvement:

Thinkers prepare the revolution; bandits carry it out.
  Mariano Azuela

Nonviolent revolution is a program of transformation of relationships, ending in peaceful transfer of power.
  Mohandas K. Gandhi

FISH Step	Activity
Focus	1 Define the present method of operation (PMO) [Process Management]
	2 Analyze the PMO to determine where to: • obliterate outdated methods • use new technology to simplify, integrate, or revolutionize existing methods
Improve	3 Implement the reengineered future method of operation (FMO)
Sustain	4 Check the stability and capability of the resulting process.
Honor	5 Act to standardize and improve the reengineered process.

© 2003 Jay Arthur

Process Reengineering
Improve and Sustain

Redesign The Process

To succeed in the turbulent 2000s, you will need both innovation (reengineering) and laser-focused improvement (kaizen). Without paradigm shifts (i.e., innovation) you cannot keep up. Without laser-focused improvement, you cannot improve on innovations quickly enough to survive. You need both! Joel Barker describes three key players in a paradigm shift:
Paradigm shifters--who discover the new paradigm
Paradigm pioneers--who figure out how to use and continuously improve (e.g., developers of Yahoo--Internet).
Paradigm settlers--who come in when it's "safe."

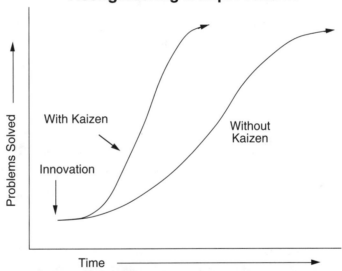

Key Tools

There are <u>three key tools</u> in the reengineering process:

- <u>Process Flowcharts</u> - to define the current and redesigned process

- <u>Matrices</u> - to <u>align</u> the new process with the customer's requirements

- <u>Control Charts</u> - to measure the performance of the process.

With these three tools you can redesign virtually any process to be tenfold better, faster, and cheaper.

Instructor Guide

Design For Six Sigma
Process

When to Use?
When you want to achieve at least 4.5-Sigma when designing new products or redesigning existing products.

While the laser-focused (FISH) improvement process deals in existing products and processes, Design for Six Sigma (DFSS) deals with new or redesigned products. The goal is to design a product and the process to produce it in ways that will achieve at least 4.5 sigma (1000 PPM).

DFSS starts with the Voice of the Customer to determine their requirements and convert them into CTQs (critical to quality measures). These requirements are converted into internal design, parts, process, and production requirements. These internal requirements are compared to internal standards and competitive benchmarks. Along the way you also seek to find and prevent every way that a part or process can fail. Then, various production factors, like time and temperature, are compared to determine the optimal settings for best results.

Process

FISH	Step	Activity
Focus	1	**Use Quality Function Deployment** (QFD) to gather the "voice of the customer" to establish requirements, benchmark the competition, and establish Critical to Quality (CTQs) design requirements.
Improve	2	**Use Pugh Concept Selection Matrix** to choose among design alternatives.
	3	**Use Failure Mode and Effects Analysis** (FMEA) to identify potential problems with products and processes and design methods to prevent failures.
Select	4	**Use Design of Experiments** (DOE) to create a "robust" design that optimizes key factors in the production process.
	5	**Use Reliability tools** like the block diagram to optimize product and process reliability.
	6.	**Use simulations** to evaluate the design.

Design for Six Sigma

Design or Redesign The Product and Process

While most design processes barely deliver 2 or 3 sigma, DFSS can help you design or redesign a product or service in ways that will deliver 4.5 sigma (1,000PPM) right from the start. And do it in half the time. Why? Because it forces you to think through all of the options and potential tar pits *before* you commit people time, and money to a new product or service. As much as 70% of your total costs can be traced back to the design process.

DFSS begins with the customer's needs, wants, and wishes (first external customers, then internal customers) and translates them into **CTQs** (Critical to Quality metrics). There are several acronyms for the DFSS process: DMADV, IDOV, DCCDI, DMEDI, but they all boil down to the same tools used in the same order

DMADV - Define and Measure requirements, Analyze options, Design the process, and Verify its performance.

IDOV - Identify customer requirements, Design best solution, Optimize design and performance, Validate its performance.

FISH - Find out what your customers want, Identify the alternative ways to give them what they want, Select the best overall solution, and Harmonize the resulting product, service, and process with ever changing requirements.

Benefits
- Reduced time to market (speed)
- Reduced design changes (defects)
- Reduced design & life cycle costs
- Increase quality
- Improved customer satisfaction.

Key Tools

There are five key tools in the reengineering process:
- Quality Function Deployment (QFD) - to translate the voice of the customer into business requirements.
- Pugh Concept Selection Matrix - to evaluate design alternatives.
- Failure Modes and Effects Analysis (FMEA) - to evaluate and prevent the most likely causes of product or process failure.
- Design of Experiments (DOE) - to evaluate various alternatives and create the most "robust" design possible.
- Block Diagrams - to increase the reliability of the product or service.

With these five tools you can redesign virtually any process to be tenfold better, faster, and cheaper.

© 2003 Jay Arthur

Instructor Guide

Quality Function Deployment
Design for Six Sigma

QFD

Always design a thing by considering it in its next larger context.
  Eliel Saarinen

Quality Function Deployment (QFD) recognizes that six sigma quality must be integrated into every aspect of a new product, process, or service. Quality cannot be added on or tested in later. QFD seeks to take the customer's requirements (the "Voice of the Customer") and turn them into design requirements that will dramatically improve customer satisfaction while slashing traditional cycle times.

QFD Process

The QFD process is a rigorous planning process to ensure that customer's requirements will be satisfied. It can slash the time required to design new products or services, and it can be used to reengineer business processes.

Phase	Step	Activity
Service	1	Gather the Voice of the Customer through surveys and analysis of customer correspondence and complaints. Benchmark against key competitors.
	2	Develop and analyze the design requirements (House of Quality).
Delivery	3	Develop a "blueprint" of the delivery process.
Materials	4	Identify the people, process, and technology needed to establish and maintain product and service delivery.
Operations	5	Act to implement process.

QFD House of Quality
Design for Six Sigma

Why?

When?

Designing new products and processes to achieve Six Sigma.

How?

1. **Service**: Gather the Voice of the Customer through surveys and analysis of customer correspondence and complaints. Develop and analyze the design requirements (House of Quality).

2. **Delivery**: Develop a "blueprint" of the delivery process.

3. **Materials** Identify the people, process, and technology needed to establish and maintain product and service delivery.

4. **Operations**: Act to implement process.

QFD (Quality Function Deployment) is a rigorous planning process to ensure that customer's requirements will be satisfied. It can slash the time required to design new products or services, and it can be used to reengineer business processes.

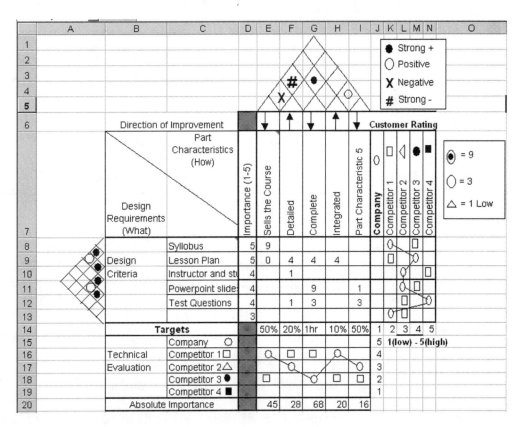

Instructor Guide

Pugh Concept Selection
Design for Six Sigma

Evaluate Design Concepts

Pugh Concept Selection is a simple matrix that compares design requirements (CTQs) against various design alternatives and ranks each against the current "baseline."

Step	Activity
1.	**Identify the criteria or requirements.**
2.	**Identify the various design alternatives.** Many companies are using TRIZ (the Theory of Inventive Problem Solving) to identify these alternatives (see below).
3.	**Rate each concept** against each criteria as better than, worse than, or the same as the baseline.
4.	Select the optimum design alternative (more +'s).

TRIZ

TRIZ began in Russia in 1946 with an assumption that there are *universal principles of innovation* and that these can be learned by anyone. Research on over 2 million patents revealed that 1) *problems and solutions are repeated across industries and sciences*, 2) *patterns of technical evolution are repeated as well*, and 3) *innovations use insights gleaned from* outside *of their field*.

There are at least 40 principles underlying TRIZ. The six overarching components of TRIZ include:

1. Setting high goals (voice of the customer)

2. Cause & Effect (identifying critical functions-CTQs)

3. Eliminate or replace harmful, corrective, enabling, or productive functions or parts (FMEA)

4. Improve function to the extreme (FMEA, DOE)

5. Resolve contradictions (FMEA, QFD, DOE)

6. Expand and consolidate

Pugh Concept Selection Matrix
Design for Six Sigma

Why?

When?
When evaluating design alternatives.

To evaluate various design alternatives against an existing baseline.

	A	B	C	D	E	F	G	H	I
					Design Concepts				
1	Pugh Concept Selection Matrix Comparison Criteria	Current Process (Baseline)	Plug and Play	Bolt and Screw		Concept 4	Concept 5	Concept 6	Concept 7
2	Faster Assembly		+	-					
3	Harder to accidentally disconnect		-	S					
4									
5	Criteria								
6	Criteria								
7									
8									
9									
10	Total +'s		1	0	0	0	0	0	0
11	Total -'s		1	1	0	0	0	0	0
12									
13	Compare current with		+ Better Alternative						
14	selected alternatives		- Worse Alternative						
15			S Same Alternative						
16									
17			Focus on alternative with the most +'s and fewest -'s						

© 2003 Jay Arthur

Six Sigma Instructor Guide

Failure Modes & Effects Analysis
FMEA

FMEA

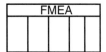

FMEAs help analyze the design of a part or assembly to: 1) identify potential failures, 2) rank these failures, and 3) find ways to eliminate these problems *before* they occur. FMEAs proactively, rather than reactively, reduce the defects, time, and cost associated with potential errors by preventing crises.

FMEA Process

Step	Activity
1.	Enter part name and/or block diagram the components
2.	List each potential failure mode
3.	Describe effects of each type of failure
4.	Rank severity of failure (see below)
5.	Classify any special characteristics
6.	List every potential cause or failure mechanism for each failure mode.
7.	Rank the likelihood of occurrence of each failure/cause
8.	List prevention/detection controls
9.	Rank detection (see below)
10.	Identify actions to reduce severity, occurrence, and detection.

Benefits

- Improves quality, reliability, and safety.
- Reduces development costs and time
- Improves customer satisfaction
- Helps select optimum design alternatives
- Identifies potential interactions
- Provides a basis for developing diagnostic and management procedures.

Severity of Effect:	Customer Experience:	Part
1. None		
2. Very Minor	Not annoyed	Fit/finish/squeak/rattle
3. Minor	Slightly annoyed	
4. Very Low	Minor annoyance	
5. Low	Some dissatisfaction	Comfort items impaired
6. Moderate	Discomfort	Comfort items inoperable
7. High	Dissatisfied,	operable but impaired
8. Very High	Very dissatisfied,	inoperable
9. Hazardous with warning	Potentially hazardous effect (gradual failure)	
10. Hazardous w/o warning	Sudden, hazardous failure	

Occurrence Rating	Sigma Level
1. Remote <.01/1000	
2. Low - 0.1/1000	5 Sigma
3. Low - 0.5/1000	
4. Moderate - 1/1000	
5. Moderate - 2/1000	
6. Moderate - 5/1000	4 Sigma
7. High - 10/1000	
8. High - 20/1000	
9. Very High 50/1000	3 Sigma
10. Very High >100/1000	2 Sigma

Likelihood of Detection:
1. Almost Certain
2. Very High
3. High
4. Moderate High
5. Moderate
6. Low
7. Very Low
8. Remote
9. Very Remote
10. Absolute Uncertainty

© 2003 Jay Arthur

FMEA
Failure Modes and Effects Analysis

When?
1. New product/design
2. Modified design
3. Existing design applied in new environment

Recognize and evaluate potential failures of a product or process; Identify actions to prevent the failure; and document the process.

Blank Template

Item/Part / Function	Potential Failure Mode	Potential Effect(s) of Failure	Sev	Class	Potential Cause(s) / Mechanism(s) of Failure	Occur	Current Design Controls Prevention	Current Design Controls Detection	Detect	R.P.N.	Recommended Action(s)	Responsibility & Target Completion Date	Actions Taken	Sev	Occur	Detect	R.P.N.
What are the Functions, Features, or Requirements?	What can go wrong? 1. No function 2. Degraded function 3. Unintended function	What are the Effects?	How bad is it?		What are the Causes?	How often does it happen?		How can this be detected and prevented?			What can be done? Changes to: 1. Design 2. Process 3. Controls 4. Documentation						

Severity of Effect: 1. None 2. Very Minor
Occurrence Rating: 1. Remote < .01/1000 2. Low - 0.1/1000
Detection: 1. Almost Certain 2. Very High
Detection: 1. Almost Certain 2. Very High
RPN= Risk Priority Number

Sample

Item/Part / Function	Potential Failure Mode	Potential Effect(s) of Failure	Sev	Class	Potential Cause(s) / Mechanism(s) of Failure	Occur	Current Design Controls Prevention	Current Design Controls Detection	Detect	R.P.N.	Recommended Action(s)	Responsibility & Target Completion Date	Actions Taken	Sev	Occur	Detect	R.P.N.
Front Door LH HBHX-0000-A	Corroded interior lower door pannels	Deteriorated life of door leading to:	7		Upper edge of protective wax application specified for inner door panels is too low	6		Vehicle general durability test veh. T-118	7	294	Add laboratory accelerateed corrosion testing	A. Tate-Body Engineering Corrosion testing	Based on test results, upper edge spec raised 125mm	7	2	2	28
		1. Unsatisfactory appearance due to rust through point over time	7		Insufficient wax thickness specified	4		Vehicle general durability testing	7	196	Conduct DOE on xax thickness	A. Tate	Thickness is adequate.	7	2	2	28
		2. Impaired function of interior door hardware	7		Insufficient room between panels for spray head access	4		Drawing evaluation of spary head access	4	112	Add team evaluation using design aid buck and spary head	Body Engineering	Evaluation showed adequate access.	7	1	1	7

© 2003 Jay Arthur — Six Sigma Instructor Guide

Instructor Guide

Failure Modes & Effects Analysis
Process FMEA

PFMEA

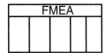

PFMEAs help analyze a <u>process</u> to: 1) identify potential failures, 2) rank these failures, and 3) find ways to eliminate these problems *before* they occur. FMEAs proactively, rather than reactively, reduce the defects, time, and cost associated with potential errors by preventing crises.

PFMEA Process

Benefits

- Assists in designing controls to reduce or prevent the production of unacceptable products
- Establishes priorities for improvement efforts
- Documents the rationale for process choices

Step	Activity
1.	Flowchart the process
2.	Describe process and function
3.	List each potential failure mode
4.	Describes effects of each type of failure
5.	Rank severity of failure
6.	Classify any special characteristics
7.	List every potential cause or failure mechanism for each failure mode.
8.	Estimate the likelihood of occurrence of each failure/cause
9.	List prevention/detection controls
10.	Rank detection
11.	Identify actions to reduce severity, occurrence, and detection.

Severity of Effect:
1. None
2. Very Minor
3. Minor
4. Very Low
5. Low
6. Moderate
7. High
8. Very High
9. Hazardous with warning
10. Hazardous w/o warning

Occurrence Rating
1. Remote <.01/1000
2. Low - 0.1/1000
3. Low - 0.5/1000
4. Moderate - 1/1000
5. Moderate - 2/1000
6. Moderate - 5/1000
7. High - 10/1000
8. High - 20/1000
9. Very High 50/1000
10. Very High >100/1000

Detection:
1. Almost Certain
2. Very High
3. High
4. Moderate High
5. Moderate
6. Low
7. Very Low
8. Remote
9. Very Remote
10. Absolute Uncertainty

Process FMEA

When?
1. New process
2. Modified process
3. Existing process applied in new environment

Recognize and evaluate potential failures of a process; Identify actions to prevent the failure; and document the process.

						Process FMEA (EMEA)		AIAG Third Edition						
Item/Process		Process Responsibility:				EMEA Number								
Subsystem		Key Date:				Page		of						
Model Years						Prepared by:								
Core Team:						EMEA Date						Action Results		
Process Function Requirements	Potential Failure Mode	Potential Effect(s) of Failure	S e v	Potential Cause(s) / Mechanism(s) of Failure	O c c	Current Process Controls Prevention	Current Process Controls Detection	D e t	R.P.N.	Recommended Action(s)	Responsibility & Target Completion Date	Actions Taken	S e v	O c c	D e t	R.P.N.
Simple process description, number, and its purpose	Describe how product or process could potentially fail.	What the internal or external customer might notice or experience: noice, impaired function.		Describe how the failure could occur described in terms of what chan be corrected or controlled:		Process methods and controls to prevent failure.	Process methods and controls to detect failure.		0	Changes to reduce severity, occurrence, and detection ratings.	Name of organization or individual and target completion date	Actions and actual completion date				0
									0							0
									0							0
									0							0
									0							0
		Severity of Effect:		**Occurrence Rating**		**Detection:**				RPN= Risk Priority Number						
		1. None		1. Remote <.01/1000		1. Very High										
		2. Very Minor		2. Low - 0.1/1000		2. Very High										
		3. Minor		3. Low - 0.5/1000		3. High										
		4. Very Low		4. Moderate - 1/1000		4. Moderately High										
		5. Low		5. Moderate - 2/1000		5. Moderate										
		6. Moderate		6. Moderate - 5/1000		6. Low										
		7. High		7. High - 10/1000		7. Very Low										
		8. Very High		8. High - 20/1000		8. Remote										
		9. Hazardous with warning		9. Very High 50/1000		9. Very Remote										
		10. Hazardous w/o warning		10. Very High >100/10		10. Almost Impossible										

Sample

						Process FMEA (EMEA)		AIAG Third Edition								
Item/Process		Process Responsibility:				EMEA Number										
Subsystem		Key Date:				Page		of								
Model Years						Prepared by:										
Core Team:						EMEA Date						Action Results				
Process Function Requirements	Potential Failure Mode	Potential Effect(s) of Failure	S e v	Potential Cause(s) / Mechanism(s) of Failure	O c c	Current Process Controls Prevention	Current Process Controls Detection	D e t	R.P.N.	Recommended Action(s)	Responsibility & Target Completion Date	Actions Taken	S e v	O c c	D e t	R.P.N.
Manual application of wax inside door	Insufficient wax coverage over specified surface	Deteriorated life of the door leading to:	7	Manually inserted spray head not inserted far enough	8	Visual check each hour for film thickness and coverage		5	280	Add positive depth stop to sprayer Automate spraying	MFG Engineering	Stop added Rejected due to complexity press limits determined and controls installed	7	2	5	#
		Unsatisfactory appearance due to rust	7	spray heads clogged	5	Test spray pattern	Visual check each hour for film thickness and coverage	5	175	Use DOE on viscosity vs temperature vs pressure	MFG Engineering		7	1	5	#
		Impaired function of interior door hardware	7	spray head deformed	2	Preventive maintenance	Visual check each hour for film thickness and coverage	5	70	None	MFG Engineering					0
			7	Spray time insufficient	8	Operator instructions and lot sampling (10 doors/shift)		7	392	Install spray timer	Maintenance	Automatic spray timer installed. Operator starts timer controls shut off.	7	1	7	#
									0							0

© 2003 Jay Arthur   Six Sigma Instructor Guide

Instructor Guide

Design of Experiments

Why?
When?
After problem solving to identify plan for implementing changes.

How?

1. Determine objectives, potential causes, and factors (usually 2, 3, or 4 factors).

2. Select experimental factors, identify potential interactions, and levels (+/-, high/low)

3. Choose appropriate design (4, 8, or 16 trials) and randomize sequence of trials

4. Run the experiment

5. Analyze the data to determine interactions and best factor levels

6. Verify results

The purpose of DOE is to quickly and efficiently discover the optimum conditions that produce top quality. Trial-and-error is the slowest method of discovering these optimal conditions and usually misses the effects of various interactions. DOE significantly reduces the time and trials necessary to discover the best combination of factors to produce the desired level of quality and robustness.

	A	B	C	D	E	F	G	H	I	J
1	Design of Experiments				L4					
2	Factor	Factor Name			Level 1		Level 2			
3	A	Die Temperature			Room temp		200 degrees			
4	B	Pour Time			6 sec		12 sec			
5	AB	Die Temperature X Pour Time								
6										
7	Design Factors				Trial Responses					
8	Trial	A	B	AB	1	2	3	Average		
9	1	-	-	+	122.3	121.5	121.9	121.90		
10	2	-	+	-	128.5	129	128.2	128.57		
11	3	+	-	-	127.3	127.9	127.8	127.67		
12	4	+	+	+	125.8	125.2	126.2	125.73		
13				Average	125.98	125.90	126.03	125.97		
14		(1)	3	2						
15	Interactions		(2)	1						
16				(3)						
17				Pour Time Low	Pour Time High					
18	Low (-)	125.23	124.78	121.90	128.57					
19	High (+)	126.70	127.15	127.67	125.73					
20										
21	Anova	Factor			df	SS	MS	F	Effect	Contrast
22	Source	Die Temperature			1	6.45	6.45333	37.9608	1.5	8.80
23		Pour Time			1	16.80	16.8033	98.8431	2.4	14.20
24		Die Temperature X Pour Time			1	55.47	55.47	326.294	-4.3	-25.80
25		Error			8	1.36	0.170			
26		Total			11	80.09				

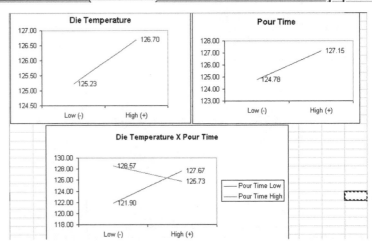

© 2003 Jay Arthur — I-104 — Six Sigma Instructor Guide

Design of Experiments
Robust Design for Six Sigma

Purpose   Experiment efficiently to create robust designs

The purpose of DOE is to quickly and efficiently discover the optimum conditions that produce top quality. Trial-and-error is the slowest method of discovering these optimal conditions and usually misses the effects of various interactions. DOE significantly reduces the time and trials necessary to discover the best combination of factors to produce the desired level of quality and robustness.

Design Factors

	A	B	AB
1	-	-	+
2	+	-	-
3	-	+	-
4	+	+	+

Many factors affect the quality of a good or service. In manufacturing, time, temperature, pressure, etc. can all affect the quality and durability of a part or product. There may also be interactions among these various factors that affect product quality and robustness. For example, the amount of time various coats of a car's finish are baked at various temperatures will affect the durability of the paint over time.

DOE can also be used in service industries although it takes a little more thought to determine the factors, their high (+) and low (-) levels, and to quantify their interactions.

DOE Process

DOE Designs
Plackett-Burnam
Full Factorials
$2^2, 2^3, 2^4$

Taguchi
Screening
L4, L8, L16, etc.

Focus	1.	Determine objectives (biggest-smallest, most-least, closest to target), potential causes, and factors (usually 2, 3, or 4 factors).
	2.	Select experimental factors, identify potential interactions, and levels (+/-,high/low)
	3.	Choose appropriate design (4, 8, or 16 trials) and randomize sequence of trials
Improve	4.	Run the experiment
	5.	Analyze the data to determine interactions and best factor levels
	6.	Verify results
Sustain	7.	Implement the optimum factors

Block Diagram
Reliability Design for Six Sigma

Why?

When?

When designing systems that need extra reliability (e.g., the space shuttle has 5 redundant computers, one is programmed separately.

To design reliability of systems. Use block diagrams for equipment and flow charts for processes. There are three types of block diagrams:

- **System** - physical relationship of major system components
- **Functional** - Categorize system components according to the function they provide.
- **Reliability** - depicts the effect of a component failure on the system's function.

How?

1. Lay out a diagram of parallel and serial components
2. Identify probability of failure of each component (value 0-1).
3. Multiply serial component probabilities (a*b)
4. Calculate parallel probabilities. (a+b-a*b)
5. Use parallel parts to increase reliability.

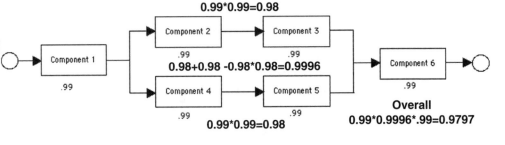

Honor Your Progress

Quality Reviews

For Six Sigma to be successful, it needs ongoing, periodic attention to recognize, review, and revise improvement efforts. Reviews conducted by all levels of management will help make the improvement effort visible and elevate its importance in the eyes of employees. It also helps management shift from reactive fire fighting to proactive fire prevention. Here's a suggested review schedule:

- Quarterly - Presidential Review
- Bimonthly - Middle Management Review
- Monthly - Line Management Review

Review Process

Step	Activity
1	Track and review improvement efforts.
	Which objectives are... take action to...
	• on target recognize participants
	• in trouble assist in achievement
	• in jeopardy refocus or revise
2	Refocus objectives, teams, and improvement efforts as required.

Recognize and Reward

Improvement efforts require recognition and reward to ensure that they continue. There are two elements to consider when evaluating team performance:

- Results - How did the business improve?
- Process - How well did the team apply Six Sigma?

Refocus and Revise

The original improvement focus, like all business objectives, will need to be adjusted as new information comes to light. Did the indicators (i.e., measurements) truly measure the customer's requirements? Were some of the improvement targets too high? Where some too low and achieved easily? Were some of the means to achieve the targets unfocused or incorrect? If so, then revise and refocus the improvement efforts. Continue this cycle forever--focus the improvement efforts, make improvements, sustain, an honor them.

Instructor Guide

Implementation Plan

Action Plan Elicit from participants what could be some of their "next steps" for implementing breakthrough improvement.

ACTION				

Six Sigma Implementation Plan

Purpose   Identify who will do what and when

Action will remove the doubt that theory cannot solve.
  Tehyi Hsieh

Implementing Six Sigma will require a careful plan to ensure that it will take root and flourish. Otherwise, it would be like scattering seeds over hardened soil and hoping for the best. First, leadership must take the time to clear the fields and develop a plan that focuses the organization on a few key areas for *laser-f*ocused improvement. This requires setting long- and short-term objectives, measures, and targets.

Then, leadership can identify the few core processes that are essential to the organization's success. Using process management, you will want to define and measure these processes. Some of these processes will be so far out of date and inflexible that only benchmarking or reengineering will close the gap between where you are and where your customers expect you to be in the near future. Problem-solving will quickly move your other processes in line with customer expectations.

Action Planning

A good Six Sigma implementation plan will identify:

- **What** activities to implement
- **How** to do them
- **Who** will do them
- **When** they will be started and completed
- **How** they will be measured

Instructor Guide

Implementation Plan

Exercise

Purpose: Develop an action plan to initiate breakthrough improvement

Agenda:
1. In groups of 3-to-5, have participants develop an action plan to initiate their improvement effort.
2. What three key actions can they take now that will start their organization on the road to breakthrough improvement?
   - Planning to focus the improvement
   - Problem solving at key leverage points
   - Process Management to stabilize key processes

Limit: 1 hour

Six Sigma Implementation Plan

ACTION PLAN

	WHAT?	HOW? (Specific Action)	WHO?	WHEN? Start Complete	MEASURE? (Results)
Focus	Gather Voice of the Customer, Business, & Employees				
	Develop Master Improvement Story				
Improve	Initiate improvement projects • Problem Solving • Core Process • Benchmarking • Reengineering				
Sustain	Track and evaluate indicators				
Honor	Reward teams Review and Refocus targets as appropriate				

© 2003 Jay Arthur

Instructor Guide

Implementing Six Sigma

Save $250,000 or more per project and add it to the bottom line.

The big mistake most companies made implementing TQM was focusing on the number of people trained and the number of teams started. When I listen to many companies talk about implementing Six Sigma, I unfortunately hear the same story: how many green and black belts trained. How many projects started. If you want to fail miserably at great expense, try implementing Six Sigma the same way most companies implemented TQM: train everyone and start teams with no focused objectives or accountability.

Six Sigma is about producing measurable results, in dollars, that you can add to the bottom line. Every company who is not actively solving these kinds of problems is wasting $40 out of ever $100 they spend. If you're a $5 million company, that's $2 million in waste and rework. You can, with focus, add $1.5 million of that back onto your bottom line. If you're a $50 million company, then you're wasting $20 million and can add $15 million back onto the bottom line. Do you want to pump up profits without selling one more customer? Start finding and preventing the waste and rework in your business.

The next few pages describe the implementation in a high-level overview. Adapt it to your organization. Get the early adopters on board. Start generating meaningful results and the rest of the organization will catch on. It may take a little longer, but it will provide bottom-line profits along the way and increase your chances of Six Sigma taking deep and permanent root in your organization.

Executive Summary

Breakthrough Improvement

Plan of Action: Implement Laser-focused improvement:
- create 6σ skills in key employees
- create measurable results during implementation
- transfer the skills of laser-focused improvement to the initial wave of team members
- transfer ongoing implementation to internal consultant-trainers selected from successful initial teams.

Process:

Implementing Breakthrough Improvement

Focus 1. Learn the essence of laser-focused improvement and process management.

Focus 2. Focus your improvement efforts to achieve six sigma reductions in cycle time, defects, and cost, which translate to dramatic improvements in customer satisfaction, speed, quality, and profitability. 4% of the business creates over 50% of the problems.

Improve 3. Employ the 6σ problem solving process and apply it in multiple parallel teams to achieve quantum leaps in improvement.

Sustain 4. Stabilize and sustain the improved processes to ensure continued high performance.

Honor 5. Develop internal consultant-trainers to continue the implementation of 6σ.

Considerations:

The statistics are ominous: over half of all TQM efforts failed; the same will be true for Six Sigma. A typical company invests 42 hours per employee per year to develop 6σ skills. At a loaded cost of $100/hr, that means $4200 per employee. If you have 100 employees, that's $420,000/yr <u>just for training</u>, add another 40 hours for team meetings. With Laser-Focused Improvement, employees get 2 hours of Just-In-Time training and 14 hours of results-creating experience. When a critical mass--20-30% of the people--have this deep experience with 6σ, the change cascades through the company.

Approach:

1. Under NO circumstances should you attempt to train everyone and do everything. As shown on the next page, leadership must focus on the top one, two, or three priorities and develop the first steps of the improvement story. If leadership, guided by skilled consultants, cannot do this, neither can a team.

2. Once you know exactly which problems to solve first, you will know who should be on the root cause analysis team. This team should meet for no more than two days to hammer out the root causes and proposed solutions (i.e., countermeasures).

3. Implementation teams should implement and sustain the process.

Instructor Guide

Implementing Six Sigma

Through trial and error I have found a proven way to succeed at these problem solving efforts every time.

Focus
1. The leadership team with a Six Sigma assistant must focus the improvement efforts by identifying the most painful problems to be solved and developing the Master 6σ Story. Until this is completed, it is difficult to identify the ideal team members for each problem solving team. I have found that if the leadership team can't do this, then no team could either. This prevents wasted team effort.

Improve
2. Teams identify the root causes of each problem. Multiple teams are required to achieve true breakthroughs—50% or greater reduction in defects, cycle time, and cost. Once they've identified the causes and countermeasures in meetings that should take no more than 1-2 days, they return to their jobs and leaders project manage the implementation of the improvements. The problem solvers may be involved if they are part of the implementation. I have found that teams should <u>never</u> meet for 1-2 hours a week until they complete their analysis. If a problem is worth solving, it's worth solving now. And any team can identify the root causes in a day or two or their problem was poorly focused from the very beginning.

Sustain
3. Once the implementation team has verified the success of the improvement, then a system of measures and processes can ensure that the improvement continues. Without defined processes and ongoing measures, the system will eventually slide back to its former level of performance.

Honor
4. Leadership needs to review and recognize root cause and implementation team members for their participation in the success of a project. Many companies tie management and Six Sigma Black Belt compensation to the success of these efforts as well.

This process will take you from 3 Sigma to 5 sigma in 24-36 months. To reach 6 Sigma, you may need to use QFD and DOE to redesign your processes.

Laser-Focused Improvement To Achieve Six Sigma

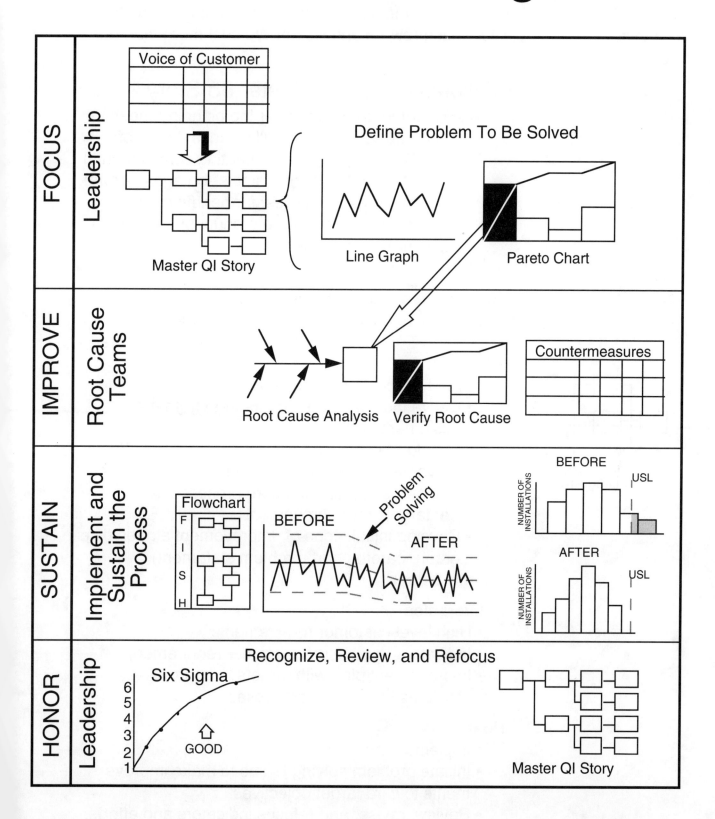

Step 1: Focus The Improvement

LifeStar
Jay Arthur

Problem:
- It takes too long! We miss commitments!
- Too many defects! Too many customer complaints
- Too much rework and waste! We cost more than our competitors.

Solution: Develop a Master 6σ Story

Prework: Gather information for the planning meeting
- What do our customers really want in terms of:
    - Speed (making and meeting commitments)
    - Quality (low or no defects--outages, errors)
    - Value (i.e., low cost for benefit received)
- What processes deliver the product or service (i.e., value) we provide?
- What measures are already in place?

Develop Laser-Focus

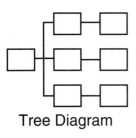

Tree Diagram

Focused, results-oriented meeting
Purpose: To develop an improvement plan and management system with indicators.

Agenda:
- Analyze customer requirements and pain
- Derive indicators (i.e., measures) from the customer's requirements
- Develop Master 6σ Story including targets for improvement based on customer requirements, current areas of pain, and core processes.
- Set targets for improvement
- Develop initial steps of improvement stories (line graph, pareto chart, problem statement)

Limit: 2 days

Deliverables
- High level customer requirements
- Indicators to measure customer requirements
- Improvement plan with targets to improve key problems in the core processes.

Post Work
- Implement measures
- Initiate problem solving teams to make improvements toward target objectives.
- Review, revise, and refocus indicators and efforts.

Step 2: Improve The Process

Create Breakthrough Improvements

Line Graph

Pareto Chart

Cause-Effect Diagram

Problem:
- It takes too long! We miss commitments!
- Too many defects! Too many customer complaints
- Too much rework and waste! We cost more than our competitors.

Solution: Focused, Intensive Problem Solving

Prework: Gather information for the meeting
- Performance data:
    - Speed (making and meeting commitments)
    - Quality (low or no defects--outages, errors)
    - Value (i.e., low cost for benefit received)
- Potential root causes

Analyze An Element Of The Master 6σ Story

Purpose: To develop focused improvement stories that will deliver laser-focused improvements in speed, quality, and profitability.

Agenda:
- Analyze value-added flow to remove waste and rework.
- Analyze customer pain
- Analyze root causes of problem
- Develop countermeasures and action plan

Limit: 1-2 days

Deliverables
- Improvement stories ready for implementation
- Project plan for implementation

Post Work--Results and Standardization

Purpose: To measure results and standardize improvements that will deliver laser-focused improvements in speed, quality, and cost.

Agenda:
- Implement countermeasures and track results
- Iterate until targets are achieved.
- Use Process Management to stabilize, standardize, and sustain the improvements.

Step 3:
Sustain The Improvement

Problem:
- customers want stable, dependable, reliable products and services and we can't provide them 100% of the time.
- problems return after improvement efforts
- unusual circumstances cause problems

Solution: Process Management

Prework: Gather information for meeting
- What do our customers really want in terms of:
  - Speed (making and meeting commitments)
  - Quality (low or no defects--outages, errors)
  - Value (i.e., low cost for benefit received)
- What processes deliver the product or service (i.e., value) we provide?
- What key areas of customer pain?

Creating A Process Management System

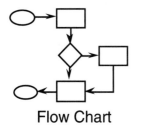
Flow Chart

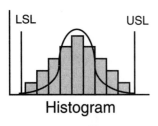
Control Chart

Develop a Process Management System
  Purpose: To identify, document, and stabilize core processes that will deliver the speed, quality, and value our customers crave.
  Agenda:
  - Flowchart the high-level process and the next lower level
  - Develop indicators that will measure and predict the process' ability to meet customer needs for speed, quality, and cost.
  - Develop project plan to implement the process and indicators.

  Limit: 2-3 days

  Deliverables
  - Process documentation--flowcharts
  - Indicators to measure process performance.
  - Project plan for improvement teams

  Post Work--Improvement teams
    Purpose: To analyze root causes and deliver laser-focused improvements in speed, quality, and cost.

Step 4:
Train Internal "Blackbelts"

Problem:
- need to train additional employees and solve key problems in a cost effective and timely manner.

Solution: 6σ "BlackBelt" Training

Prework: Experience all aspects of the 6σ process (i.e., serve on successful teams)
- Improvement Focus
    - Speed (making and meeting commitments)
    - Quality (low or no defects--outages, errors)
    - Value (i.e., low cost for benefit received)
- Problem Solving--root cause analysis
- Process Management--stabilization

Create Internal Experts

Develop Accelerated Learning And Training Skills
Purpose: To train internal consultants to maximize skill transfer and accelerate organizational success.
Agenda:
- Training flow: Show-Do-Know
- Laser Focus demonstration, elicitation, instructional exercise, and feedback.
- Problem Solving demonstration, elicitation, instructional exercise, and feedback.
- Process Management demonstration, elicitation, instructional exercise, and feedback.

Limit: 2 days

Deliverables
- Instructor skills
- Training plans

Post Work--Certification Course
Purpose: To observe instructors presenting in-house course. Recommend areas for improvement. Certify in 6σ methodology.
Agenda:
- Instructors train/facilitate team in 6σ methodology.
- Feedback and areas for improvement
- Certification

Limit: 2 days

Instructor Guide

Other Processes and Tools

Affinity Diagram

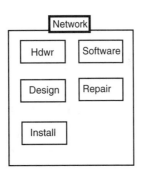 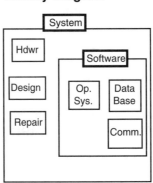 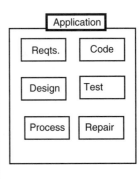

Systems Diagrams

Force Field

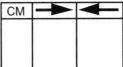

Tree Diagram

Cost of Quality

Step	Costs	Total
		Benefit

Instructor Guide

General Planning
Affinity Diagram

Affinity Diagram

The affinity diagram helps break old patterns of thought, reveal new patterns, and generate more creative ways of thinking. The affinity diagram helps organize the team's thoughts most effectively when: the issues seem too large and complex; you need to break out of old, traditional ways of thinking; everything seems chaotic; or there are many customer's requirements.

Parts of the overall problem will naturally cluster into components that can be investigated easily. The process is simple:

Affinity Diagram Process

Step	Activity
1.	**State the issue** to be examined in broad terms: "What are the issues surrounding or involving . . . • the delivery of very low defect products or services • reducing cycle time • reducing waste or rework
2.	**Generate and record ideas** using Post-it™ notes. Begin sticking them on a wall or large sheet of easel paper where everyone can see them. Ensure that everyone is included. Ask for a "headline" to describe each thought. Note the contributor's initials.
3.	**Arrange the cards in related groupings**. As you generate ideas, the person at the board may begin grouping the available notes as they are offered and keep the intensity of note generation going as long as possible.
4.	**Complete the groupings.** Involve the group in clustering the notes into 6-10 related groupings. Have everyone stand and do this silently. Be prepared for some "loners." Avoid forcing them into a group. Some notes may need to be duplicated for different groupings.
5.	**Choose a word or phrase that captures the intent of each group** and place it at the top as a header card. If there isn't one already, then create one with a word or phrase that does capture the intent.

© 2003 Jay Arthur

Instructor Guide

Cost of Quality Analysis

When?
During laser-focus of improvement effort. (Is this problem worth solving?)

Evaluate the true costs of a defect or error. This is the foundation of making a business case for the change.

How?

1. Identify each step in the Fix-it process.
2. Assign a time in minutes to each task and loaded rate.
3. Identify any material costs associated with each step.
4. Identify any external costs of this failure.
5. Identify an lost opportunity, asset, or business costs.
6. Set a target for reducing the error (e.g., 50%)
7. Estimate the total cost of achieving this level of prevention.
8. Evaluate ROI and payback period.

Cost of Quality Worksheet

Problem Description: Service Order Errors **Type:** Internal

	Tasks	Average Hours/Task	Hourly Rate	Cost of Task	Material Costs	External Failure Cost	Total Cost of Non-Conformance
4	1. Analyze Service Order Error	0.17	$60	$10.00	$3.00	$0.00	$13.00
5	2. Fix Error	0.08	$60	$5.00	$3.00	$0.00	$8.00
6	3. Admin	0.05	$60	$3.00	$0.00	$0.00	$3.00
7	4. Billing Costs Due to Error	0.03	$60	$2.00		$0.00	$2.00
8	**Total Cost Per Failure**						$26.00
9	Service Order Errors/year						221,000
10	1. Lost Opportunity Costs					$0.00	$0.00
11	2. Lost Assets Costs					$0.00	$0.00
12	3. Lost Business Costs					$0.00	$0.00
13	Additional Failure Costs						$0.00
14	**Annual Failure Cost**						$5,746,000.00
16	Basic tasks to fix the problem	Average min/60	Loaded rate	Calculated cost	Expenses	Customer or Employee found	Total

Return on Investment and Payback

Target Reduction	50%		$2,873,000
Prevention Costs			$225,000
ROI			$13:$1
Payback Period (days)			17

© 2003 Jay Arthur   Six Sigma Instructor Guide

Instructor Guide

Force Field Analysis

Exercise (Optional)

Purpose: Develop a force field analysis to determine potential barriers to implementation.

Agenda:

- In small groups, have participants select one proposed countermeasure.
- Have participants identify the forces (people, process, machine, materials, environment, budget) that support or impede the implementation.
- Have participants identify any additional action items required to remove the barriers to implementation.

Limit: 30 minutes

Purpose: Define Problem

Agenda:
- Pareto Chart
- Problem statement

Limit: 30 minutes

Counter measures	Forces	
	For	Against
Increase time spent with customers	Customer loyalty and lifetime value.	Cost per customer

© 2003 Jay Arthur — Six Sigma Instructor Guide

Instructor Guide
Systems Diagrams

Purpose — Evaluate the <u>circular</u> cause-effects in the existing system

Systems Diagram Symbols

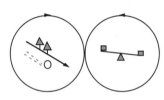

It is best to do things systematically, since we are only human, and disorder is our worst enemy.
  -Hesiod

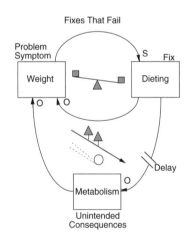

While root cause analysis is the ideal way to handle "linear" cause effects, not everything in a company is linear. Companies are "living systems" that scientists call "complex adaptive systems." What they mean by this is that companies have "non-linear" cause-effects. Through feedback (like a microphone), small causes can be amplified into huge effects.

The systems diagram helps you map these non-linear (i.e., circular) cause-effects between the related items. The systems diagram shows the cause-effect relationships among many key elements. It can be used to identify the causes of problems or to work backward from a desired outcome to identify all of the causal factors that would need to exist to ensure the achievement of an outcome.

A systems diagram uses a few simple symbols to show the circular cause-effects in a system. The symbols are:

	Indicator	Measures the effect of some force in the system (amount or number of ...)
	Arrow	Showing the cause-effects among indicators.
	Delay	Lag between cause and effect
	S or O	Cause Effect relationship **S**ame- A↑ B↑ **O**pposite- A↑ B↓

© 2003 Jay Arthur   Six Sigma Instructor Guide

Instructor Guide

Systems Diagrams

Systems Diagram Process

Use Post-it™ notes for both the indicators and the arrows. This way, the system will remain easy to change until you have it clearly and totally defined.

Begin by brainstorming things that are either increasing or decreasing (e.g., amount of turnover, number of calls, number of orders, number of items ordered, etc.).

Step	Activity
1.	**State the system, problem or issue** under discussion
2.	**Generate cause-effect issues** by brainstorming things that are either increasing or decreasing (e.g., amount of turnover, number of calls, number of orders, number of items ordered, etc.). Note: There should be more than nine and less than fifty notes when completed, otherwise the problem is either too simple or too complex for this method.)
3.	**Draw <u>one-way</u> arrows to indicate the cause-effect relationship** among all of the components of the diagram. Avoid two way arrows; decide which component has the most influence and draw the arrow in one direction only.
4.	**Identify the effects.** If the an *increase* in A causes B to increase, then put the letter "S" on the arrow near B to indicate that they move in the same direction. If an *increase* in A causes a *decrease* in B, then put the letter "O" on the arrow near B to indicate that they move in opposite directions. (e.g., an increase in the amount of downsizing causes a corresponding decrease in morale and productivity.)

Instructor Guide

Intermediate Planning
Tree Diagram

Tree Diagram

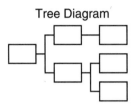

The tree diagram can map specific tasks to primary and secondary goals. It maps the methods required to achieve corporate goals. The tree diagram shows the key goals, their sub-goals, and key tasks. It can help identify the sequence of tasks or functions required to accomplish an objective. The tree diagram can help translate customer desires into product characteristics. It can also be used like an Ishikawa diagram to uncover the causes of a particular problem.

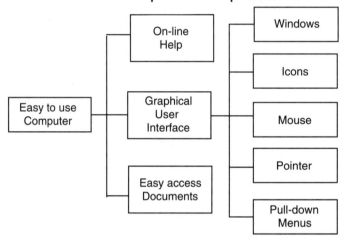

Tree Diagram Process

The process for constructing the tree diagram is similar to the Ishikawa diagram in many respects:

Step	Activity
1.	**Develop a clear statement of the problem**, issue, or objective to be addressed. Place it on the left side of a board, wall, or easel and work toward the right.
2.	**Brainstorm all of the sub-goals, tasks, or criteria** necessary to accomplish or resolve the issue.
3.	**Repeat this process** using each of the sub-goals until only actionable tasks or elements remain.
4.	**Check the logic of the diagram** in the same way as the Ishikawa: Start at the right and work your way back to the left by asking: "If we do this, will it lead to the accomplishment of the previous task?"

Instructor Guide

Quality Reading List

Brassard, M., *The Memory Jogger Plus+*™, GOAL/QPC, 1989.
(The seven management and planning tools)

Brefogle, Forrest, Implementing Six Sigma, John Wiley & Sons, 1999.

Guaspari, John, I'll know it when I see it!, AMACOM, 1984.

Harrington, James, The Improvement Process, ASQC, 1989.

Juran, J. M., *Quality Control Handbook (5th)*, 1998.

King, Bob, *Better Designs in Half the Time*, GOAL/QPC, 1989.
(Quality Function Deployment—QFD)

Leonard, George, *Mastery*, Penguin, 1991.

Montgomery, Douglas, Introduction to SPC (4th), John Wiley & Sons, 2001.

Montgomery, Douglas, Design and Analysis of Experiments (5th), John Wiley & Sons, 2001.

Rogers, Everett, The Diffusion of Innovations, 1995.

Senge, Peter, The Fifth Discipline Fieldbook, Doubleday, 1992.

Instructor Guide

Tools Quick Reference

Action Plan	66, 104
Affinity Diagram	115
Block Diagram	105
c Chart	90
Cause-Effect Diagram	60, 94
Control Charts	81, 87
Cost-of Quality	116
Countermeasures matrix	62
Customer Requirements Matrix	40-43
Design of Experiments	104
Fishbone Diagram	60, 94
FMEA	102
Flowchart	49-51, 74
Force-field Analysis	118
Histogram	84-86, 119
Ishikawa Diagram	60, 94
Line Graph	76
Matrix Diagram	56, 120
np chart	88
p chart	88
Pareto Chart	58, 64
Pugh Concept Selection Matrix	101
QFD	100
6σ Story	69
Systems Diagrams	124
Tree diagram	44, 126
Voice of the Customer	37
u Chart	90
XR charts	92

Six Sigma Simplified

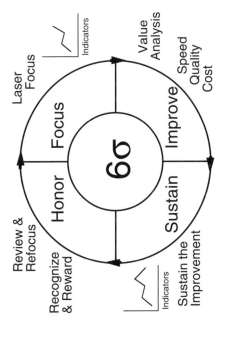

Quick Reference Card

© 2001 Jay Arthur

LifeStar
2244 S. Olive St.
Denver, CO 80224-2518

(888) 468-1537 or **(303) 753-9355**
(888) 468-1536 or (303) 753-9675 (fax)
lifestar@rmi.net
http://www.qimacros.com

Six Sigma Processes

There are three key processes in Six Sigma (see back for more detail):

Focus Like A Laser

<u>Purpose</u>: To focus the effort and achieve breakthrough improvements in speed, quality, and cost. Using the Voice of the Customer, develop a "Master 6σ Story" that links and aligns multiple teams and improvement efforts to achieve quantum leaps in performance improvement. 4% of your business causes over 50% of the costs. Prevent those problems and add $250,000 or more per project back onto the bottom line.

Improve Speed, Quality, Cost

<u>Purpose</u>: To improve customer satisfaction by identifying and eliminating the root causes of problems involving time, defects, or cost. Also known as: quality improvement, root cause analysis, or continuous improvement; this process uses data to analyze problems and eliminate their root causes.

Sustain the Improvement

<u>Purpose</u>: To define and stabilize any process. Also known a SPC--statistical process control, this process uses data to evaluate the ability of any business process to predictably and consistently meet the customer's requirements. It serves as a basis to systematically improve any process and maintain the gains from such improvements.

Benchmarking to adopt best practices.

Reengineering with QFD to revolutionize existing processes with the aid of technology.

Key Tools

Tree Diagram: Systematically link ideas, targets, objectives, goals, or activities in greater and greater detail. It shows key goals, sub goals, measures, and tasks required to accomplish an objective.

Matrix: Compare two or more groups of ideas, determine relationships among the elements, and make decisions. It helps prioritize tasks or issues to aid decision making and shows linkages between large groups of characteristics, functions, and tasks.

Line Graph: Show data trends over time. The Y-axis (left) shows the defects, time, cost and the X-axis (bottom) shows time (minute, hour, day, week, etc).

Pareto Chart: Focus the improvement effort by identifying the 20% (vital few) of the contributors that create 80% of the time delay, defects, or costs in any process.

Cause-Effect: Analyze the root causes of problems It begins with major causes and works back to root causes.

Flowchart: Show the flow of work through a process including all activities, decisions, and measurement points.

Control Chart: Help analyze, sustain, and monitor the current levels of process stability and to identify key issues for problem solving or root cause analysis.

Histogram: Determine the <u>capability</u> (i.e., the level of performance the customers can consistently expect) of the process and the distribution of measurable data.

© 2003 Jay Arthur

Six Sigma Instructor Guide

Laser Focus

Purpose: Focus the improvement effort to avoid wasting valuable time and money.

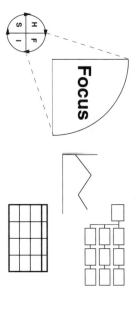

Key Tools
- Tree diagram
- Line graph
- Matrix diagram

Process

FISH	Step	Activity
Focus	1	Use the Voice of the Customer to develop a Master Improvement Story.
	2	Identify and track the indicators
	3	Set targets for improvement.
Improve	4	Initiate process improvements
Sustain	5	Sustain the improvements
Honor	6	Honor your progress
	7	Review and refocus objectives, teams, and improvement efforts as required.

Improve Speed, Quality, Cost

Purpose: Make Breakthrough improvements in Speed, Quality, and Cost

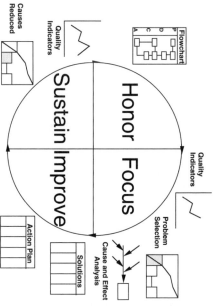

Key Tools
- Line graph
- Pareto Chart
- Ishikawa (i.e., Fishbone) Diagram

Process

FISH	Step	Activity
Focus	1	Define the problem: Reduce delay, defects, or cost
	2	Analyze the problem
Improve	3	Implement the countermeasures
Sustain	4	Stabilize to lock in the improvements
Honor	5	Review, recognize, and refocus

Sustain the Improvement

Purpose: Stabilize and Sustain the Improvement

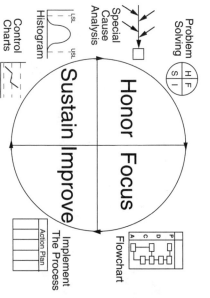

Key Tools
- Flow Chart
- Control Chart (stability)
- Histogram (capability)

Process

FISH	Step	Activity
Focus	1	Refine the process
	2	Identify the "quality" and "process" indicators
Improve	3	Implement the process and indicators
Sustain	4	Check the process for stability and capability
Honor	5	Review, recognize, and refocus Continue Improvement

Six Sigma Instructor Guide

© 2003 Jay Arthur

Affordable SPC Software for Six Sigma and ISO

*Using the 6σ Macros you can **automatically** draw all of these graphs, charts, and tools using Microsoft® Excel (PC or Macintosh).*

The QI Macros save our clients an estimated 75% of their documentation time and cost. Joe Farina - PerformTech

You've thought of everything. I used 12 features to complete a Six Sigma project in just a day. The QI Macros aren't sexy, just practical! John Weisz

Accelerate Your Success: The QI Macros automate all of the charts, graphs, and diagrams you need. With the macros, simply select the data you want to graph and click on the pull-down menu; the QI Macros will do all the math and draw the graph for you.

Save Time: Even if you don't know the first thing about Microsoft Excel, you can start drawing professional-looking control charts, histograms, and other critical charts and diagrams in a matter of minutes.

Avoid Effort: Your data is probably already in Excel. Unlike other SPC packages, the QI Macros work directly off your Excel data. You don't have to copy and paste your Excel data into another program or a different Excel format.

Increase Profit: You don't need to know the math and statistics behind SPC, DOE, or GageR&R to immediately start analyzing every aspect of your business and making breakthrough improvements in productivity and profitability.

Save Money: Unlike the Cadillac of SPC software which is great for statisticians, but is hard to use and costs over $1,000, **the QI Macros are only $129**, and there are discounts when you buy five or more copies. You also get flowcharts and fishbones that leverage Excel's drawing capability. Other QI Macros templates automate Cost of Quality, Gage R&R, QFD, and DOE. (**If you had to buy all of these packages independently they could easily cost $2,000 or more**, but they wouldn't fit together seamlessly in an Excel workbook.) With what you save you can afford to give these tools to more people to make even greater breakthroughs.

Gain Consistency: Share your charts and graphs with anyone who has Excel...they don't have to have the QI Macros to read the charts! Leverage your investment!

Who is using the QI Macros? 1,000s of hospitals, manufacturers, and service companies. With the QI Macros, paretos, histograms, and control charts are just a mouse click away. All you have to do is select the data you want graphed and, in just seconds, *the macro will do all the math and draw the graph* for you. They also include over 35 easy-to-use, fill-in-the-blanks templates for all your other Six Sigma needs:

    DFSS: QFD, DOE, Pugh Concept Selection, FMEA, EMEA
    Focus: Voice of the Customer, Tree diagram
    Improve: Line graph, pareto chart, and Ishikawa diagram
    Sustain (SPC): Flowcharts, control charts, histograms
        Measurement Systems Analysis: Gage R&R
        18 charts and 35 templates to automate all of your documentation.

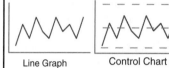
Line Graph   Control Chart

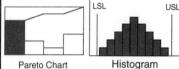
Pareto Chart   Histogram

Cause-Effect Diagram   Bar Chart

QFD Matrix   Pie Chart

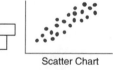
Flow Chart   Scatter Chart

Tree Diagram

Download a FREE Pareto Chart Maker from qimacros.com/freestuff.html

36 pg. User Guide + 3 1/2" Floppy Disk

Requirements: PC or Mac running **Excel Version 5.0, Office 95, 97, 2000, & XP.** 1Mb is all the storage you need.

Supports all Windows 3, 95, 98, 2000, Me, NT, XP. Single user and network compatible! **90 day money back guarantee!**

LifeStar, 2244 S. Olive St. Denver, CO 80224 • (888) 468-1535 or (303) 281-9063

© 2003 Jay Arthur   Six Sigma Instructor Guide

Six Sigma On $5 A Day!

Increase Your Profits by $250,000 or More!

Using A Simple, Easy-to-Use, Six Sigma System

$435.95 + $14 S&H

The Six Sigma books are a non-intimidating approach to systematic improvement that retain all the goodness of Six Sigma without watering it down.

– **Don Poskin, GTE Communications**

Turn your processes into profits!

As little as 4% of your business processes produce over 50% of your waste and rework. Through laser-focused improvement, you can cut these costs of waste and rework by 20-35% of every dollar spent!

Warning! Over half (50%) of all TQM efforts failed. That's about a 1-Sigma performance. Pathetic! The costs of these failures can run into the millions for wasted training and teams. Unfortunately, most companies are using the same, dumb implementation strategy for Six Sigma! Don't let your Six Sigma implementation fail. Six Sigma Simplified delivers a proven, research-based method to implement Six Sigma (or anything else for that matter) in ways that ensure success with substantially lower risk. The secret? Harness the power of "diffusion" to "crawl-walk-run" your way to success.

Only you can decide if you really want to turn your cash cow into a golden goose. If so, you need the Six Sigma System. Haven't you waited long enough to start making your business competition-proof?

For Leaders, Managers, and Quality Professionals

If you're an operational manager with responsibilities for managing the delivery of a product or service, couldn't you use an extra $250,000 in cost reductions or profitability? Wouldn't you like to shift some of your "Fix-it" Factory workers back onto the mainline? With the complete system, you can figure out exactly where to focus your efforts for maximum return.

Here's What You Get in the Complete System: (Item #290)

- Six Sigma Simplified (128 pgs) for team members and training participants (Item #205)
- Six Sigma Instructor's Guide (192 pgs) for Money Belt team leaders and trainers (Item #205)
- Six Sigma Simplified Training Video and Student Guide (Item # 265, 2 tapes, 120 minutes)
- Six Sigma Simplified Audio (Item #225, 4 cassettes, 240 minutes)
- QI Macros add-in software for Excel to automate your documentation (Item #230)
- QI Macros Computer-based Training CD-ROM (Item #237)

Find out more at: www.qimacros.com/sixsig290.html

To: Executives, Managers, and Quality Professionals   **From:** Jay Arthur
Re: Getting Results with Six Sigma

Over the last decade, I have had the opportunity to work with companies from healthcare to aerospace. I've learned from failure and success how to make Six Sigma take root and grow in organizations. The difference is a focus on results, and taking a crawl-walk-run approach to implementation. Unfortunately, I see too many companies using the teams and training strategy of TQM to implement Six Sigma. <u>Over half of these are doomed to failure.</u>

I've found that it is possible to create breakthrough improvements every time. I've helped companies save millions of dollars, and shave weeks and months off sloth-like processes. And so can you. And you can do it with a minimum investment that will generate maximum results.

To Order: Call 888-468-1535   Fax 888-468-1536

© 2003 Jay Arthur   Six Sigma Instructor Guide

Money Belt Team Leader-Trainer System (Item #260)

Want to do your own classroom and team training? Professors in colleges and blackbelts in companies are using the Six Sigma Simplified System to train teams and greenbelts. Once you know the secret, you can use the accelerated learning process built into Six Sigma Simplified to *train employees in as little as 2 hours*. Then *solve problems in a day or less* and *create bottom-line results while you develop employee skills*. You'll help team members learn how to make million dollar improvements through the use of:

- Six Sigma Instructor's Guide (192 pgs, see pg. 10)
- Six Sigma Simplified (128 pgs, see below)
- QI Macros to automate your documentation (pg. 2)
- QI Macros Computer-based Training CD-ROM (pg. 10)

$199.95 + $10 S&H

Six Sigma Starter Kit (Item #255)

For those who don't immediately need the instructor guide, we created a starter package:
- Six Sigma Simplified (128 pgs) for team members and greenbelts
- **QI Macros for Six Sigma data analysis**
- QI Macros Computer-based Training CD-ROM

$169.95 + $10 S&H

Six Sigma Simplified Workbook - Greenbelt Training Made Easy (#205)

Team leaders: Do your team members have trouble understanding how to use the tools of Six Sigma in a realistic way that creates results quickly? Do you have trouble deciding what measurements to use and how to identify mission-critical problems? Do you have trouble finding root causes?

Consider the Six Sigma Simplified workbook for team member training. It shows you how to achieve $1,000,000 improvements using a proven, results-oriented methodology:

1. Uses a story about Robin Hood to teach the improvement process
2. Gives a basic overview of Six Sigma's benefits and key components
3. Explains each step in the improvement process and the associated tools

$27.95 + $8 S&H 128 pages

- *Focus* on the 4% of your business where dramatic improvements are not only possible, but necessary.
- *Improve* using core tools like line graphs, pareto charts, and root cause analysis.
- *Sustain* the improvement using flowcharts, control charts, and histograms (otherwise you'll backslide).
- *Honor* your progress using simple review and recognition processes to ensure continued success.

4. Provides simple, fill-in-the-blanks forms and templates for each tool with examples
5. Explains how to choose and use control charts to sustain the improvement

Six Sigma Tools Covered (find out more at www.qimacros.com/sixsigma.html)

Focus	Improve	Sustain	DFSS (Overview)
Tree Diagram	Line graph	X Control Charts	Quality Function Deployment
Voice of the Customer	Pareto Chart	c, np, p, u Control Charts	Design of Experiments
7 Planning Tools	Ishikawa/Fishbone	Histograms	
Benchmarking	Countermeasures Matrix	Flow Chart	
	Action Plan		

If you *still* don't think you can afford $400 to save $250,000, download our FREE Six Sigma Starter Guide, QI Macros Demo, and User Guide from www.qimacros.com!

Order Online: www.qimacros.com

© 2003 Jay Arthur

Yes! I want Jay Arthur's fast, fun and easy-to-use Six Sigma Simplified, Million Dollar Money Belt System to work for me! *Please send the software, training material, audio and video indicated below. Offer good until 6/30/03.*

☐ **Complete Six Sigma System: (#290)**
QI Macros (#230)+Training CD (#237)
90 pg. Six Sigma Tools (#239)
192 pg. Instructor Guide (#210)
128 pg. Six Sigma Simplified (#205)
Video (#265) & Audio (#225)
Only $449.95 Save $45
includes $14 S&H, add $23 for FedEx

☐ **Team Leader System: (#260)**
QI Macros (#230)+Training CD (#237)
90 pg. Six Sigma Tools (#239)
128 pg. Six Sigma Simplified (#205)
192 pg. Instructor Guide (#210)
Only $209.95 Save $35
includes $10 S&H, add $25 for FedEx

☐ **Six Sigma Starter Kit: (#255)**
QI Macros (#230) +Training CD (#237)
90 pg. Six Sigma Tools (#239)
128 pg. Six Sigma Simplified (#205)

Only $179.95 Save $25
includes $10 S&H, add $25 for FedEx

Order Form

Qty.	Item	Individual Items	Price	FedEx	3-Day Mail	Item Total
	230	QI Macros for Excel (20% discount for 5 or more)	$129	$20	$8	
	237	QI Macros Training CD-ROM	$19.95	$20	$8	
	239	Six Sigma Tools (Example Book)	$19.95	$20	$8	
	205	Six Sigma Simplified Team Member Workbook	$27.95	$20	$8	
	210	Six Sigma Instructor Guide–Greenbelt Training Made Easy	$39.95	$20	$8	
	225	Six Sigma Simplified (4-Audiotapes & Training Guide)	$49.95	$20	$8	
	265	Six Sigma Simplified (2-Videotapes & Training Guide)	$199.95	$20	$8	
	Shipping and Handling: $8+$2 (Mail) or $20+$5 (Airmail/FedEx) for each additional item					
					Order Total	

Please type or print clearly or attach business card here

Your Name _____
Company _____
Mailing Address _____
P.O. Box _____ Apt/Ste. _____
City, ST, Zip _____
Phone _____
Fax _____
Sign up for our Free Six Sigma Ezine:
Email _____

Yes! We also accept Purchase Orders!

Purchase Order Number _____
(to prevent duplicate shipments, <u>never</u> send confirming POs)

☐ VISA ☐ MasterCard ☐ AMEX
_____ Exp._____
Signature _____

☐ I've enclosed my check, VISA, MasterCard, or AmEx.
☐ I want to try them out **Absolutely Risk Free**. Please send my order immediately. *I have 30 days to pay the invoice or return them with no obligation.*
Special Bonus #1 Six Sigma Money Belt Action Plan
Special Bonus #2: Six Sigma Quick Reference Card
☐ *For a FREE analysis of what Six Sigma can do for your bottom line, call 888-468-1537 today.*

Orders Only
(To minimize errors please complete this form and fax your order)

FAX your order toll-free to: (888) 468-1536 or (303) 753-9675

Mail: LifeStar, 2244 S. Olive St. Denver, CO 80224-2518

Orders-only, Call Toll-free: (888) 468-1535 or (Please have your item # ready) (303) 281-9063

Questions about the QI Macros?
8a.m. and 5p.m. MST (888) 468-1537 or (303) 756-9144 or
email: knowwareman@qimacros.com

90 Day, Unconditional, No Risk, Money-back Guarantee